TEC

MATHEMATICS EXERCISES

Level I

.Prod.E.
Lecturer
chnology

STANLEY THORNES (PUBLISHERS) LTD.

First published in 1978 by Stanley Thornes (Publishers) Ltd.
EDUCA House, 32 Malmesbury Road, Kingsditch
CHELTENHAM, GL51 9PL

ISBN 0 85950 485 9

Typeset by the Alden Press (London and Northampton) Ltd.
and printed in Great Britain at The Pitman Press, Bath

CONTENTS

Section A

MANIPULATION OF NUMBERS AND CALCULATIONS

When you have completed this section you should be able to:

1. Apply the four basic arithmetic operations to expressions involving integers, fractions and decimals.
2. Solve problems relating to ratio, proportion and percentage.
3. Apply the precedence rules to multiple calculations involving fractions and decimals.
4. Use and convert denary numbers to and from standard form.
5. Use and convert binary numbers to and from denary.
6. Ensure that answers to numerical problems are reasonable.
7. Use 4 figure tables of square, square root and reciprocals.
8. Use 4 figure $\log_{10}$ tables.
9. Use a slide rule for simple calculations.
10. Perform basic arithmetic operations on a calculating machine.

1. Multiply the following using longhand:
   a) 5×374
   b) 5629×9
   c) $17\,135 \times 11$
   d) 13×2914
   e) 27×457
   f) 153×259

2. Divide the following using long division:
   a) $1554 \div 6$
   b) $3045 \div 7$
   c) $7059 \div 13$
   d) $3174 \div 23$
   e) $41\,496 \div 28$
   f) $23\,343 \div 31$

3. Find the L.C.M. of the following numbers:
   a) 2, 3, 4, 8
   b) 2, 4, 5, 6, 8, 10
   c) 3, 7, 9, 14, 21
   d) 1, 2, 3, 4, 5, 6, 7
   e) 9, 11, 18, 27, 44
   f) 13, 39, 65, 143

4. Find by the method of prime factors the H.C.F. of the following numbers:
   a) 84 and 96
   b) 190 and 798
   c) 24, 42 and 51
   d) 78, 182 and 195
   e) 84, 140 and 616
   f) 2117 and 4745

5. Reduce the following fractions to their lowest terms:
   a) $\frac{25}{100}$
   b) $\frac{6}{42}$
   c) $\frac{7}{56}$
   d) $\frac{12}{16}$
   e) $\frac{12}{32}$
   f) $\frac{14}{64}$
   g) $\frac{9}{15}$
   h) $\frac{11}{121}$
   i) $\frac{18}{27}$
   j) $\frac{48}{144}$
   k) $\frac{121}{22}$
   l) $\frac{221}{238}$

6. Multiply or divide the following fractions:
   a) $\frac{3}{5} \times \frac{2}{7}$
   b) $\frac{3}{4} \times \frac{6}{7}$
   c) $\frac{4}{9} \times \frac{9}{10}$
   d) $2\frac{1}{3} \times \frac{6}{7}$
   e) $3\frac{3}{4} \times 1\frac{1}{3}$
   f) $7\frac{4}{11} \times 3$
   g) $\frac{1}{2} \div 3$
   h) $\frac{7}{12} \div \frac{14}{15}$
   i) $4 \div 1\frac{3}{5}$
   j) $\frac{3}{14} \div \frac{3}{7}$
   k) $1\frac{2}{7} \div \frac{3}{5}$
   l) $\frac{11}{26} \div 1\frac{9}{13}$

7. Add or subtract the following fractions:
   a) $\frac{2}{3} + \frac{1}{2}$
   b) $\frac{4}{7} + \frac{7}{8}$
   c) $\frac{5}{11} + \frac{11}{12}$
   d) $\frac{6}{19} + \frac{20}{57}$
   e) $\frac{5}{6} + \frac{9}{14}$
   f) $\frac{13}{15} + \frac{15}{17}$
   g) $\frac{3}{4} - \frac{1}{2}$
   h) $\frac{7}{8} - \frac{5}{6}$
   i) $\frac{5}{12} - \frac{1}{6}$
   j) $\frac{9}{11} - \frac{3}{7}$
   k) $\frac{4}{5} - \frac{3}{15}$
   l) $4\frac{6}{7} - 1\frac{1}{6}$

1. Express as improper fractions:

a) $2\frac{1}{8}$ c) $3\frac{1}{16}$ e) $1\frac{5}{12}$ g) $3\frac{5}{64}$
b) $5\frac{3}{4}$ d) $2\frac{1}{5}$ f) $4\frac{5}{32}$ h) $3\frac{1}{7}$

2. Reduce to whole or mixed numbers:

a) $\frac{163}{4}$ c) $\frac{127}{2}$ e) $\frac{121}{64}$ g) $\frac{23}{7}$
b) $\frac{37}{16}$ d) $\frac{12}{4}$ f) $\frac{21}{8}$ h) $\frac{125}{25}$

3. Reduce to their lowest terms:

a) $\frac{12}{64}$ c) $\frac{144}{60}$ e) $\frac{8}{20}$ g) $\frac{24}{144}$
b) $\frac{10}{32}$ d) $\frac{12}{48}$ f) $\frac{40}{64}$ h) $\frac{64}{1000}$

4. Find the value of the following:

a) $\frac{1}{2} + \frac{1}{8} + \frac{5}{12}$ d) $2\frac{1}{4} + 5\frac{3}{8} + 1\frac{13}{16}$ g) $8\frac{1}{8} - 4\frac{3}{16}$ j) $\frac{65}{100} \times 5\frac{5}{8} \div 4\frac{1}{2}$
b) $1\frac{1}{3} + \frac{5}{16} + \frac{9}{48}$ e) $\frac{2}{3} - \frac{1}{8}$ h) $4\frac{17}{64} - 2\frac{11}{32}$
c) $\frac{2}{3} + \frac{5}{9} + \frac{7}{18}$ f) $4\frac{1}{7} - 3\frac{4}{5}$ i) $\dfrac{4\frac{1}{2} \times 6\frac{3}{4}}{1\frac{4}{5}}$

5. a) If the load a vehicle will carry has a mass of $2\frac{1}{2}$ tonne and this load is $\frac{5}{9}$ of the vehicle's mass, what is the total mass of the vehicle plus load?

b) A tank containing 40 litres of oil leaks at the rate of $1\frac{1}{4}$ litres per hour.
How many hours will be taken to empty the tank?

c) A vehicle travels a distance of 64 km and uses $9\frac{1}{2}$ litres of petrol. Calculate the fuel consumption in km/litre.

d) A machine component costs £1.25. If the material accounts for $\frac{7}{20}$ and labour $\frac{19}{40}$, the remainder being profit, what is the amount of profit?

e) A housewife wishing to check her greengrocery bill needs to calculate $1\frac{1}{2}$ kg @ 12 p/kg, $\frac{3}{4}$ kg @ 16 p/kg, 2 packets weighing 3 kg @ 14 p/kg and $1\frac{1}{4}$ kg @ 24 p/kg. What is the total cost?

Express the following *ratios* as *fractions* in their lowest terms:

1. 4 m to $2\frac{1}{2}$ m
2. 6p to £1
3. $\frac{3}{20}$ to $\frac{1}{6}$ litre
4. 18 : 6
5. 1 cm to 1 m
6. $4\frac{1}{2}$ to 6
7. $1\frac{1}{4}$ kg to 5 kg
8. A craftsman is paid £1.95 per hour and his apprentice at £0.95 per hour.
   What is the ratio of the apprentice's wage to the craftsman's wage?
9. A map scale is 25 mm to 1500 m.
   What is the ratio between distances on the map to actual distances on the ground?
10. A railway line rises 1250 mm in a distance of 1 km.
    Give this slope as a ratio of rise to distance.
11. A tapered bar with a 100 mm long taper has a minor diameter of 3.68 mm and a major diameter of 9.16 mm.
    What is the taper per mm?
12. 8 cm^3 of mercury has a mass of 108.8 g.
    What is the relative density of mercury? (1 cm^3 of water has a mass of 1 g).
13. The ratio of the cost of labour to the cost of material in decorating a room is 4 : 3.
    What is the cost of materials if the total cost is £56?
14. A tinsmith reckons that the ratio of the area of sheet metal used in a job to the area of waste is 13 : 2.
    What area is wasted from 10 sheets of metal each $4\frac{1}{2}$ m^2 in area?
15. A machine costing £5000 when new is valued now at £4750.
    Determine the fractional depreciation.
16. A plank of wood 6 m long costs £1.35.
    Find the cost of two planks, each 5 m long, at the same rate.
17. A car consumes 4.5 litres of petrol during a run of 48 km.
    How much will it consume on a run of 200 km under similar conditions?
18. A photograph is to be enlarged from 74 × 52 mm to an enlargement making the longer side 297 mm.
    What will be the length of the other side?

1. A brass casting consists of 2 parts copper and 1 part zinc by mass. Calculate the mass of copper and zinc in a casting weighing 240 kg.

2. A solution for pickling castings is made of sulphuric acid and water in the ratio of 4.5 water to 1 acid.
How much water is needed if $2\frac{1}{2}$ litres of acid is used?

3. The ratio of the length of the con-rod to crank in an engine is $4\frac{1}{2}/1$. Find the length of the con-rod if the crank is 100 mm.

4. How much copper is required to be melted with 60 kg of zinc so as to make an alloy consisting of copper and zinc in the ratio of 7/3?

5. The mass of a casting is reduced by machining in the ratio of $\frac{1}{8}/1$. If after machining it has a mass of 80 kg, calculate the mass of the casting.

6. A motor car has a mass of 900 kg and carries 5 persons who have a total mass of 330 kg.
What fraction is the load of the total mass?

7. $\frac{3}{5}$ of the 300 employees at a factory are skilled men, $\frac{1}{4}$ are unskilled and the remainder semi-skilled.
How many semi-skilled employees are there?

8. Glass contains by mass, $\frac{3}{10}$ silica, $\frac{1}{20}$ potash and the remainder being lime.
How many kg of each are there in 160 kg of glass?

9. Three men are awarded £270 to be shared out in the ratio 2:3:4. Find out how much each man will receive.

10. A factory working full time, 5 days per week, 8 hours per day is reduced to a $3\frac{1}{2}$ day week. If fuel bills are normally £872 per week calculate how much should be saved.

11. A model $\frac{2}{7}$ full size is 360 mm long by 270 mm wide, what are the proportions of the full size article?

12. A length of wire 18 m long has a resistance of 270 ohms. Find the resistance of a length of wire 1.25 km long.

1. Find the value of the following:

a) $3 + 5 - 4$

b) $7 - 5 + 3$

c) $2 - 4 + 7$

d) $3 \times 2 + 4$

e) $5 - 2 \times 2$

f) $2 + 7 \times 3$

g) $6 - 4 \div 2$

h) $8 \div 2 + 2$

i) $5 \times 3 + 2 \times 4$

j) $3 \times (2 + 3) \times 2$

k) $11 - 3 \times (2 + 6)$

l) $5 - 3 \times 2 + 4 \div 2$

2. Evaluate the following:

a) $3 \times \frac{3}{8} - \frac{1}{8}$

b) $2\frac{1}{2} - \frac{2}{3} \times 1\frac{1}{2}$

c) $\frac{1}{2}$ of $\frac{3}{5} + \frac{1}{10}$

d) $(\frac{3}{8} - \frac{1}{4}) \times \frac{2}{3}$

e) $5\frac{1}{4} - 3\frac{1}{3} \times \frac{2}{5}$

f) $1\frac{1}{8} \times 4 + 2\frac{5}{6} \times 3$

g) $9 \div 3 + \frac{1}{2} \div 3$

h) $2\frac{1}{4} \times (3 + 1\frac{1}{3}) \div 2$

i) $(2 - \frac{3}{5}) \div (3 - \frac{1}{5})$

j) $8 + \frac{1}{2} \times 2 \div 16$

3. Find the value of the following:

a) $\dfrac{3 \times 5}{6}$

b) $\dfrac{5 - 2}{10}$

c) $\dfrac{3 - 2 \times 6}{9}$

d) $\dfrac{6 \times (2 + 5)}{4}$

e) $\dfrac{2 + 5(6 - 3)}{3 + 7}$

f) $\dfrac{\frac{1}{2} + \frac{1}{3}}{6}$

g) $\dfrac{4 \times \frac{3}{4} + \frac{1}{4}}{\frac{1}{3}}$

h) $\dfrac{\frac{1}{2}}{1\frac{1}{2} + \frac{2}{3}}$

i) $\dfrac{4\frac{1}{3} - \frac{1}{2} \times \frac{2}{3}}{\frac{9}{10}}$

j) $\dfrac{2 - \frac{3}{5}}{1\frac{5}{7} \times \frac{1}{6}}$

1. Express the following vulgar fractions in decimal form:

   a) $\frac{1}{2}$, $\frac{1}{4}$, $\frac{1}{8}$, $\frac{1}{40}$, $\frac{1}{800}$, $\frac{1}{2000}$, $\frac{1}{10\,000}$

   b) $\frac{3}{80}$, $\frac{7}{16}$, $\frac{15}{32}$, $\frac{3}{64}$, $\frac{9}{64}$, $\frac{15}{64}$

   c) $\frac{4}{25}$, $\frac{8}{125}$, $\frac{39}{500}$, $\frac{9}{4}$, $\frac{13}{8}$, $\frac{33}{16}$

2. Find in their lowest terms the vulgar fractions equivalent to the following decimals:

   a) 0.1, 0.01, 0.001, 0.5, 0.05, 0.005

   b) 0.25, 0.025, 0.75, 0.075, 0.125, 0.375

   c) 0.625, 0.875, 0.9375, 0.0625, 0.1875, 0.4375

3. Express $\frac{1}{7}$ to 6 decimal places
   Express $\frac{1}{7}$ to 5 decimal places
   Express $\frac{1}{7}$ to 3 decimal places
   Express $\frac{1}{7}$ to 2 decimal places

4. Express the following to the number of decimal places stated:

   a) 10.5743 (2) c) 0.037 45 (3) e) 2.899 (1)
   b) 2.0275 (2) d) 2.750 85 (4) f) 3.04 (1)

5. Express the following decimals to the number of significant figures stated:

   a) 0.002 741 (3) d) 11.056 (4) g) 11.056 (3)
   b) 11.056 (2) e) 25.2 (4) h) 61.204 (4)
   c) 51.73 (3) f) 61.85 (3) i) 67.75 (3)

6. What is the value of:

   a) $23.061 - 21.062$ c) $5.42 + 0.7 - 4.08$
   b) $23.97 - 11.4 + 6.03 - 7.8$ d) $5.63 - 2.12 + 7.96$

7. Evaluate the following stating your answer to the significant figures stated:

   a) 0.5×2.5 (2) h) $26.4 \div 10$ (2)
   b) 0.75×0.5 (1) i) $3.8 \div 1000$ (1)
   c) 2.05×1.75 (4) j) $2.56 \div 1600$ (2)
   d) 0.0500×0.2 (2) k) $0.0084 \div 17$ (3)
   e) 2.64×3.01 (3) l) $4.516 \div 2.51$ (3)
   f) 2.35×0.04 (1) m) $\dfrac{0.75 + 0.125 - 0.1875}{0.03}$ (4)
   g) $3.75 \div 27$ (2) n) $\dfrac{2.5 \times 0.04}{1.8}$ (3)

1. Reduce to decimal form (3 sig. fig.):
   a) $\frac{14}{19}$ b) $\frac{11}{230}$ c) $\frac{16}{3100}$

2. Express as a single decimal:
$$\tfrac{1}{2} + \tfrac{1}{4} + \tfrac{1}{8} - \tfrac{1}{16} - \tfrac{1}{32} - \tfrac{1}{64}$$

3. Find the following quotients correct to 3 significant figures:
   a) $0.034\,75 \div 150$ b) $0.4785 \div 83.1$ c) $287.3 \div 0.0045$

4. Simplify the following:
$$\frac{\tfrac{3}{8} + 0.5 - 0.875 + 2}{1 - (\tfrac{1}{2} + \tfrac{3}{4} \text{ of } \tfrac{1}{4})}$$

5. $$\frac{2.875 - 0.234 + 0.025}{1 - (\tfrac{5}{16} - \tfrac{3}{8} \times \tfrac{3}{4})}$$

6. $$\frac{0.3 + 0.68 \times 0.103}{0.912 \times 6 - 3 \times 3.175}$$

7. The boiling point of water rises 0.37 C degrees for an increase of 1 cm in the height of the barometer. Water boils at 100 °C when the barometric pressure is 76 cm.
   Find the boiling point when the barometric pressure is 78.35 cm. (Give the answer to 2 decimal places.)

8. The mass of a steel bar of length 222.25 mm is 3.29 kg.
   Find the mass of a 25 mm length of bar.

9. The electrical resistance of a certain type of wire is 1.24 ohm per cm Calculate to 2 decimal places the resistance of a piece of wire 18.02 cm long.

10. A certain engine normally develops 32 kW; by increasing the compression ratio the power is raised to 36 kW.
    Express this increase in power as a decimal.

11. An assembly is held together using 17 nuts, bolts and washers. If the nuts cost 50 p per dozen, the bolts cost 7 p each, the washers cost £1.65 per hundred, the other components a total of £6.27 and the labour involved costs £1.15, find the total cost of the assembly.

1. Express as percentages:

   a) $\frac{3}{5}$ b) 0.55 c) $\dfrac{0.45}{0.75}$ d) $\dfrac{\frac{1}{10}}{8}$

2. Evaluate:

   a) 0.15% of 1 kg c) 85% of 4500 W

   b) 2½% of £100 d) 4% of 254

3. A machine costing £6000 when new, is valued now at £3500. Determine the % depreciation.

4. Box A contains 240 screws, and box B contains 25% more than A.
   What percentage is the contents of box A compared to the number in Box B?

5. A bar of 'Babbit' metal consists of 2 parts antimony, 3 parts copper and 20 parts tin. Express these as percentages and find the mass of each in 225 kg of the metal.

6. If 32 men from a shift of 576 were absent, calculate:
   a) the percentage absent b) the percentage present.

7. A shaft weighing 22.5 N has been turned from a bar weighing 25 N.
   What percentage of the original weight was lost in turning?

8. 840 kg of latex lost 2½% of its mass on being cleaned.
   What was the mass of clean latex obtained?

9. A rod 1.5 m long is drawn out to $3\frac{1}{2}$ times its length.
   What is the percentage increase in length?

10. During a tensile test a steel wire increases in length from 290 mm to 295 mm.
    Determine the percentage elongation.

11. A 10 mm diameter steel bar is reduced to 8.5 mm diameter during a tensile test.
    Determine the percentage reduction in area.

12. The workers in a factory consist of 235 men, 171 women and 29 teenagers.
    What is the percentage of teenagers?

13. A company employing 753 people needs to reduce its staff by 15%.
    How many people will lose their jobs?

Express the following numbers in standard form:

1. a) 3107 f) 0.0107 k) 3763 p) 40 000
   b) 20 g) 0.01 l) 0.672 q) $\frac{1}{40\,000}$
   c) 56 712 h) 0.001 m) 0.801 r) $\frac{7}{1\,000\,000}$
   d) 96.3 i) 1001 n) 0.073 s) $\frac{654}{1275}$
   e) 54.23 j) 302 o) 9499 t) $\frac{0.007\,61}{0.000\,45}$

2. Express in standard form the number of kilometres travelled by a light wave in one year. (Speed of light = 300 000 km/s approx.)

3. Convert the following into normal decimal form.
   a) 7.1×10^{-4} d) 1.933×10^{-1} g) 94×10^{6}
   b) 8.34×10^{2} e) 2.347×10^{3} h) 1.0×10^{-5}
   c) 1.67×10^{3} f) 17.26×10^{-3}

4. Evaluate the following, expressing your answer in standard form.
   a) $(4 \times 10^{3}) + (3 \times 10^{3})$ h) $(9.23 \times 10^{6}) \times (1.91 \times 10^{5})$
   b) $(5 \times 10^{2}) + (6 \times 10^{2})$ i) $\frac{7.88 \times 10^{4}}{3.94 \times 10^{2}}$
   c) $(5 \times 10^{-2}) + (6 \times 10^{-2})$ j) $\frac{2.62 \times 10^{6}}{1.31 \times 10^{7}}$
   d) $(30 \times 10^{3}) + (2 \times 10^{2})$ k) $3.96 \times 10^{6} \times 1.32 \times 10^{-2}$
   e) $5 \times 10^{2} \times 3 \times 10^{4}$ l) $\frac{3.6 \times 10^{2} \times 8.4 \times 10^{3}}{1.8 \times 10^{3} \times 1.2 \times 10^{4}}$
   f) $(6.7 \times 10^{2}) \times (7.3 \times 10)$ m) $\frac{4.8 \times 10^{-2} \times 6 \times 10^{4}}{2.4 \times 10^{-3} \times 1.5 \times 10^{3}}$
   g) $(7.92 \times 10^{4}) \times (6.31 \times 10^{4})$ n) $\frac{3.9 \times 10^{-2} \times 1.6 \times 10^{4} \times 7.5 \times 10^{6}}{1.3 \times 10^{3} \times 3.2 \times 10^{5} \times 1.875 \times 10^{4}}$

5. Express the answers to the following questions in standard form.
   a) Reduce to metres: 16.43 kilometres, 50 450 cm, 75 mm
   b) Reduce to centimetres: 90 m, 0.434 km, 15 mm
   c) Reduce to km: 7500 m, 36 000 cm
   d) Express in grammes: 19.5 kg, 1750 mg
   e) Express in kg: 650 g, 254 300 mg
   f) Reduce 1.63 litres to cm^3, and 22 400 cm^3 to litres

6. Evaluate the following, expressing your answer in standard form:
   a) $\frac{7.2 \times 10^{6} - 60 \times 10^{4}}{0.12 \times 10^{3}}$ b) $\frac{0.36 \times 10^{2} + 4.8 \times 10^{3}}{42 \times 10^{2} - 1.2 \times 10^{2}}$

1. Convert to binary from denary notation:

a) 7	d) 45	g) 85
b) 11	e) 78	h) 103
c) 32	f) 61	i) 121

2. Convert to denary from binary notation:

a) 1011	d) 1011010	g) 10010010
b) 110011	e) 101011	h) 10101010
c) 10101	f) 1110111	i) 11011011

3. Add or subtract the following binary numbers:

a) 11 + 101	f) 110 − 100
b) 101 + 1101	g) 1101 − 1001
c) 11011 + 11011	h) 110011 − 101101
d) 10100 + 10011	i) 100010 − 10001
e) 111001 + 1011101	j) 1000000 − 101011

4. Convert the answers to question **3** into denary form.

a) Using four figure tables find the answer to the following:

1. 8^2
2. 9^2
3. 1.5^2
4. 1.6^2
5. 2.5^2
6. 25^2
7. 7.31^2
8. 9.613^2
9. 10.13^2
10. 7.634^2
11. 9.98^2
12. 3.01^2
13. 5.13^2
14. 6.072^2
15. 7.003^2
16. 101^2
17. 12.72^2
18. 169^2
19. 2051^2
20. $(1.762 \times 10^2)^2$
21. $\sqrt{121}$
22. $\sqrt{64}$
23. $\sqrt{49}$
24. $\sqrt{6.25}$
25. $\sqrt{2.96}$
26. $\sqrt{3.96}$
27. $\sqrt{39.6}$
28. $\sqrt{396}$
29. $\sqrt{3960}$
30. $\sqrt{10}$
31. $\sqrt{100}$
32. $\sqrt{1000}$
33. $\sqrt{10\,000}$
34. $\sqrt{59}$
35. $\sqrt{601}$
36. $\sqrt{8420}$
37. $\sqrt{8.3}$
38. $\sqrt{0.534}$
39. $\sqrt{0.064}$
40. $\sqrt{0.0036}$
41. $\dfrac{1}{0.5}$
42. $\dfrac{1}{0.25}$
43. $\dfrac{1}{5.61}$
44. $\dfrac{1}{7.81}$
45. $\dfrac{1}{96.1}$
46. 3.921^{-1}
47. 25.4^{-1}
48. 1.332^{-1}
49. 0.792^{-1}
50. 0.6341^{-1}
51. $0.098\,75^{-1}$
52. $\dfrac{1}{(3.25)^2}$
53. $\dfrac{1}{\sqrt{0.44}}$
54. $3.29^{-0.5}$
55. 5.71^{-2}

b) Evaluate using square, square root and reciprocal tables only:

1. $3.4^2 + 2.35^2$
2. $\sqrt{(2.71)^2 + (31.4)^2}$
3. $\dfrac{1}{\sqrt{(3.15)^2 - (2.92)^2}}$
4. $\dfrac{3}{\sqrt{57.2} + (0.015)^2}$
5. $\dfrac{0.02}{\sqrt{0.015} + \sqrt{0.0015}}$

1. Find the $\log_{10}$ of the following numbers:

a) 57.23	**e)** 9092	**i)** 0.100	**m)** 0.3921
b) 60.36	**f)** 7.312	**j)** 0.3012	**n)** 0.0567
c) 10	**g)** 431.6	**k)** 0.004 65	**o)** 0.000 36
d) 1000	**h)** 1.000	**l)** 0.9932	**p)** 0.006 72

2. Add the following pairs of numbers:

1	1	2	−1	−2	3	−2	1	−2	−1
2	−1	−1	2	−1	−2	−2	−3	2	−3

3. Repeat Question **2** subtracting the lower number from the top instead of adding.

4. Add the following pairs of bar numbers:

$\bar{2}.4$	$\bar{1}.5$	0.6	2.5	2.1	0.8	$\bar{2}.6$	$\bar{2}.5$	$\bar{1}.8$	0.7	1.7	1.6
1.3	3.3	$\bar{2}.2$	$\bar{3}.3$	0.8	1.7	2.7	1.5	4.7	$\bar{1}.5$	$\bar{1}.3$	$\bar{3}.5$

5. Repeat Question **4** subtracting the lower number from the top instead of adding.

6. Evaluate the following using logarithm tables:

a) $6.34 \times 2.593 \times 21.38$

b) $\dfrac{253.5 \times 16.6}{8.784}$

c) $0.336 \times 0.072\,65$

d) $\dfrac{16.8}{17.6 \times 38.4 \times 0.513}$

e) $\sqrt[5]{865.4}$

f) $6.45 \times \sqrt{27}$

g) $\sqrt{(3.142)^2 + (0.8754)^2}$

h) $165.3 \div 18.18$

i) $\dfrac{364.2}{9.28 \times 18.269}$

j) $\dfrac{100}{1654}$

k) $\sqrt[3]{96.628}$

l) $\sqrt{\dfrac{762.3 \times 1.427}{18.34}}$

m) $\sqrt{0.1964}$

n) $\sqrt[3]{0.067\,28}$

The following questions may be used to give practice in:

(1) Estimating approximate answers.
(2) Calculating using tables or slide rule.
(3) Calculating using a calculator.
(4) Limiting an answer to a reasonable number of significant figures.

They also:

(5) Illustrate the relative advantages of different methods of calculation.
(*Calculations* 11 and 12, and *Algebra* 10, 11 and 12 can also be used.)

1. If $l = l_0(1 + \alpha t)$ find l when $l_0 = 374$, $\alpha = 11 \times 10^{-6}$, $t = 13.5$
2. If $p = aW + b$ find p when $a = 0.45$, $W = 57$, $b = 2.4$
3. If $N = \dfrac{K\sqrt{T}}{L}$ find N when $K = 3.49$, $T = 0.0345$, $L = 27.1$
4. If $I = \dfrac{M}{4}(a^2 + b^2)$ find I when $M = 47.51$, $a = 0.0971$, $b = 0.0135$
5. If $I = \dfrac{BD^3}{12}$, find I when $B = 5.75$ and $D = 12$
6. If $F = \dfrac{4\pi^2 EI}{L^2}$, find F when $\pi = 3.142$, $E = 30 \times 10^6$, $I = 5.7$ and $L = 64$
   (Give the answer in standard form)
7. If $V = \dfrac{4\pi R^3}{3}$, find V when $\pi = 3.142$ and $R = 0.815$
8. If $y = \dfrac{Fa^2 b^2}{3EIL}$, find y when $F = 4480$, $a = 36$, $b = 84$, $E = 30 \times 10^6$, $I = 192$ and $L = 120$
9. If $s = \dfrac{y^2 - u^2}{2a}$, find s when $y = 88$, $u = 22$ and $a = 1.135$
10. If $R_1 = \dfrac{RR_2}{R_2 - R}$, find R_1 when $R = 10.54$ and $R_2 = 16.23$
11. If $D = \sqrt[3]{\dfrac{6V}{\pi}}$, and $\pi = 3.142$, find D when:
    a) $V = 10.4$ b) $V = 0.472$
12. If $V = c\sqrt{mi}$, find V when $c = 120$, $m = 1.508$ and $i = 3.2 \times 10^{-3}$
13. If $Q = 3.09\,BH^2$ and $B = 3.5$, find the value of Q when:
    a) $H = 1.414$ b) $H = 0.507$
14. If $L = \dfrac{1}{C(2\pi f)^2}$, find the value of L when $C = 0.4 \times 10^{-6}$, $\pi = 3.142$ and $f = 24 \times 10^3$

1. Evaluate and simplify the following fractions:

a) $2\frac{1}{3} \times 3\frac{1}{7}$ b) $1\frac{2}{3} \div \frac{15}{21}$ c) $\frac{5}{7} + \frac{3}{5}$ d) $\frac{5}{6} - \frac{4}{9}$

2. Evaluate by longhand methods:

a) 31×284 b) $4158 \div 27$ c) $1434.027 \div 23$

3. The composition of a coil of brass wire consists of 6 parts copper to 4 parts zinc by weight.
How much zinc is contained in a coil weighing 270 kg?

4. Three men took 7 days to paint the hull of a boat.
How long would it have taken 5 men?

5. Find the value of the following:

a) $3 + 5 \times 2 - 8$ b) $2\frac{1}{4} \times (3 + 1\frac{1}{3}) - 2$ c) $\dfrac{\frac{1}{3}}{\frac{2}{3} - \frac{1}{4}}$

6. a) Convert to a decimal $\frac{31}{16}$

b) Convert to a vulgar fraction 0.906 25

7. A bracket weighing 35.4 N was machined from a forging weighing 43.6 N.
Calculate the percentage waste.

8. Evaluate and then convert to normal decimal form:

$$(2.3 \times 10^5) \times (4.5 \times 10^{-8})$$

9. a) Convert 59 to binary notation

b) Convert 1 0 1 1 1 0 0 to denary form

c) 1 0 0 1 0 0 1 − 1 0 1 1 1 1

10. Evaluate using 4 figure tables:

a) $(0.0973)^2$ b) $\sqrt{374.1}$ c) $\dfrac{1}{0.0372}$ d) $\dfrac{2}{\sqrt{0.0135}}$

11. Evaluate using 4 figure log tables:

$$0.374 \times \sqrt[3]{0.0537}$$

12. If $s = \dfrac{x^2 - y^2}{2a}$ find s

a) by a rough estimate   b) by tables   c) by calculator

when $x = 34.2$, $y = 12.9$ and $a = 0.985$.

CALCULATIONS 1

1. **a)** 1870 **b)** 50 661 **c)** 188 485 **d)** 37 882 **e)** 12 339 **f)** 39 627

2. **a)** 259 **b)** 435 **c)** 543 **d)** 138 **e)** 1482 **f)** 753

3. **a)** 24 **b)** 120 **c)** 126 **d)** 420 **e)** 1188 **f)** 2145

4. **a)** 12 **b)** 38 **c)** 3 **d)** 13 **e)** 28 **f)** 73

5. **a)** $\frac{1}{4}$ **b)** $\frac{1}{7}$ **c)** $\frac{1}{8}$ **d)** $\frac{3}{4}$ **e)** $\frac{3}{8}$ **f)** $\frac{7}{32}$ **g)** $\frac{3}{5}$ **h)** $\frac{1}{11}$
i) $\frac{2}{3}$ **j)** $\frac{1}{3}$ **k)** $5\frac{1}{2}$ **l)** $\frac{13}{14}$

6. **a)** $\frac{6}{35}$ **b)** $\frac{9}{14}$ **c)** $\frac{2}{5}$ **d)** 2 **e)** 5 **f)** $22\frac{1}{11}$ **g)** $\frac{1}{6}$ **h)** $\frac{5}{8}$
i) $2\frac{1}{2}$ **j)** $\frac{1}{2}$ **k)** $2\frac{1}{7}$ **l)** $\frac{1}{4}$

7. **a)** $1\frac{1}{6}$ **b)** $1\frac{25}{56}$ **c)** $1\frac{49}{132}$ **d)** $\frac{2}{3}$ **e)** $1\frac{10}{11}$ **f)** $1\frac{191}{255}$ **g)** $\frac{1}{4}$ **h)** $\frac{1}{24}$
i) $\frac{1}{4}$ **j)** $\frac{30}{77}$ **k)** $\frac{3}{5}$ **l)** $3\frac{29}{42}$

CALCULATIONS 2

1. **a)** $\frac{17}{8}$ **b)** $\frac{23}{4}$ **c)** $\frac{49}{16}$ **d)** $\frac{11}{5}$ **e)** $\frac{17}{12}$ **f)** $\frac{133}{32}$ **g)** $\frac{197}{64}$ **h)** $\frac{22}{7}$

2. **a)** $40\frac{3}{4}$ **b)** $2\frac{5}{16}$ **c)** $63\frac{1}{2}$ **d)** 3 **e)** $1\frac{57}{64}$ **f)** $2\frac{5}{8}$ **g)** $3\frac{2}{7}$ **h)** 5

3. **a)** $\frac{3}{16}$ **b)** $\frac{5}{16}$ **c)** $2\frac{2}{5}$ **d)** $\frac{1}{4}$ **e)** $\frac{2}{5}$ **f)** $\frac{5}{8}$ **g)** $\frac{1}{6}$ **h)** $\frac{8}{125}$

4. **a)** $1\frac{1}{24}$ **b)** $1\frac{5}{6}$ **c)** $1\frac{11}{18}$ **d)** $9\frac{7}{16}$ **e)** $\frac{13}{24}$ **f)** $\frac{12}{35}$ **g)** $3\frac{15}{16}$ **h)** $1\frac{59}{64}$ **i)** $16\frac{7}{8}$ **j)** $\frac{13}{16}$

5. **a)** 7 tonnes **b)** 32 hours **c)** $6\frac{14}{19}$ km/litre **d)** $21\frac{7}{8}$p **e)** £1.44

CALCULATIONS 3

1. $\frac{8}{5}$ **2.** $\frac{3}{50}$ **3.** $\frac{9}{10}$ **4.** $\frac{3}{1}$

5. $\frac{1}{100}$ **6.** $\frac{3}{4}$ **7.** $\frac{1}{4}$ **8.** $\frac{19}{39}$

9. 1:60 000 **10.** 1:800 **11.** 0.0548 **12.** 13.6

13. £24 **14.** 6m^2 **15.** $\frac{1}{20}$ **16.** £2.25

17. 18.75ℓ **18.** 208.7 mm

CALCULATIONS 4

1. 160/80 kg **2.** $11\frac{1}{4}$ litre **3.** 450 mm **4.** 140 kg

5. 90 kg **6.** $\frac{11}{41}$ **7.** 45 **8.** 48/8/104

9. £60/90/120 **10.** £261.6 **11.** 1260/945 **12.** 18.75 kΩ

CALCULATIONS 5

1. **a)** 4 **b)** 5 **c)** 5 **d)** 10 **e)** 1 **f)** 23 **g)** 4 **h)** 6
i) 23 **j)** 30 **k)** −13 **l)** 1

2. **a)** 1 **b)** $1\frac{1}{2}$ **c)** $\frac{2}{5}$ **d)** $\frac{1}{12}$ **e)** $3\frac{11}{12}$ **f)** 13 **g)** $3\frac{1}{6}$ **h)** $4\frac{7}{8}$
i) $\frac{1}{2}$ **j)** $8\frac{1}{16}$

3. **a)** $2\frac{1}{2}$ **b)** $\frac{3}{10}$ **c)** −1 **d)** $10\frac{1}{2}$ **e)** $1\frac{7}{10}$ **f)** $\frac{5}{36}$ **g)** $9\frac{3}{4}$ **h)** $\frac{3}{13}$
i) $4\frac{4}{9}$ **j)** $4\frac{9}{10}$

CALCULATIONS 6

1. **a)** 0.5, 0.25, 0.125, 0.025, 0.00125, 0.0005, 0.0001
b) 0.0375 0.4375, 0.46875, 0.468 75, 0.140 625, 0.234 375
c) 0.16, 0.064, 0.078, 2.25, 1.625, 2.0625

2. **a)** $\frac{1}{10}$, $\frac{1}{100}$, $\frac{1}{1000}$, $\frac{1}{2}$, $\frac{1}{20}$, $\frac{1}{200}$,
b) $\frac{1}{4}$, $\frac{1}{40}$, $\frac{3}{4}$, $\frac{3}{40}$, $\frac{1}{8}$, $\frac{3}{8}$
c) $\frac{5}{8}$, $\frac{7}{8}$, $\frac{15}{16}$, $\frac{1}{16}$, $\frac{3}{16}$, $\frac{7}{16}$

3. 0.142 857, 0.142 86, 0.143, 0.14

4. **a)** 10.57 **b)** 2.03 **c)** 0.037 **d)** 2.7509 **e)** 2.9 **f)** 3.0

5. **a)** 002 74 **b)** 11 **c)** 51.7 **d)** 11.06 **e)** 25.20 **f)** 61.9 **g)** 11.1
h) 61.20 **i)** 67.8

6. **a)** 1.999 **b)** 10.80 **c)** 2.04 **d)** 11.47

7. **a)** 1.3 **b)** 0.4 **c)** 3.588 **d)** 0.010 **e)** 7.95 **f)** 0.09
g) 0.14 **h)** 2.6 **i)** 0.004 **j)** 0.0016 **k)** 0.000 494
l) 1.80 **m)** 22.92 **n)** 0.0556

CALCULATIONS 7

1. a) 0.737 b) 0.0478 c) 0.005 16 **2.** 0.765 625 **3.** a) 0.000 232 b) 0.005 76 c) 63 800 **4.** −6.4 **5.** 2.752 **6.** −0.0913 **7.** 100.87 °C **8.** 0.37 kg **9.** 22.34 ohms **10.** 0.125 **11.** £9.60

CALCULATIONS 8

1. a) 60% b) 55% c) 60% d) $1\frac{1}{4}$% **2.** a) 0.15 kg b) £2.50 c) 3825 W d) 10.16 **3.** 41.7% **4.** 80% **5.** 8%, 12%, 80%, 18 kg, 27 kg, 180 kg **6.** a) 5.56% b) 94.44% **7.** 10% **8.** 819 kg **9.** 250% **10.** 1.724% **11.** 27.75% **12.** 6.67% **13.** 113

CALCULATIONS 9

1. a) 3.107×10^3 b) 2×10^1 c) 5.6712×10^4 d) 9.63×10^1 e) 5.423×10^1 f) 1.07×10^{-2} g) 1×10^{-2} h) 1×10^{-3} i) 1.001×10^3 j) 3.02×10^2 k) 3.763×10^3 l) 6.72×10^{-1} m) 8.01×10^{-1} n) 7.3×10^{-2} o) 9.499×10^3 p) 4×10^4 q) 2.5×10^{-5} r) 7×10^{-6} s) $\dfrac{6.54 \times 10^2}{1.275 \times 10^3}$ t) $\dfrac{7.61 \times 10^{-3}}{4.5 \times 10^{-4}}$

2. 9.46×10^{12}

3. a) 0.000 71 b) 834 c) 1670 d) 0.1933 e) 2347 f) 0.017 26 g) 94 000 000 h) 0.000 01

4. a) 7×10^3 b) 1.1×10^3 c) 1.1×10^{-1} d) 3.02×10^4 e) 1.5×10^7 f) 4.891×10^4 g) $4.997\,52 \times 10^9$ h) $1.762\,93 \times 10^{12}$ i) 2×10^2 j) 2×10^{-1} k) 5.2272×10^4 l) 1.4×10^{-1} m) 8×10^2 n) 6×10^{-4}

5. a) 1.643×10^4, 5.045×10^2, 7.5×10^{-2} b) 9×10^3, 4.34×10^4, 1.5, c) 7.5, 3.6×10^{-1} d) 1.95×10^4, 1.75 e) 6.5×10^{-1}, 2.543×10^{-1} f) 1.63×10^3, 2.24×10^1

6. a) 5.5×10^4 b) 1.1853

CALCULATIONS 10

1. a) 1 1 1 b) 1 0 1 1 c) 1 0 0 0 0 0 d) 1 0 1 1 0 1 e) 1 0 0 1 1 1 0 f) 1 1 1 1 0 1 g) 1 0 1 0 1 0 1 h) 1 1 0 0 1 1 1 i) 1 1 1 1 0 0 1

2. a) 11 b) 51 c) 21 d) 90 e) 43 f) 119 g) 146 h) 170 i) 219

3. a) 1 0 0 0 b) 1 0 0 1 0 c) 1 1 0 1 1 0 d) 1 0 0 1 1 1 e) 1 0 0 1 0 1 1 0 f) 1 0 g) 1 0 0 h) 1 1 0 i) 1 0 0 0 1 j) 1 0 1 0 1

4. a) 8 b) 18 c) 54 d) 39 e) 150 f) 2 g) 4 h) 6 i) 17 j) 21

CALCULATIONS 11

a) **1.** 64 **2.** 81 **3.** 2.25 **4.** 2.56
5. 6.25 **6.** 625 **7.** 53.44 **8.** 92.41
9. 102.6 **10.** 58.28 **11.** 99.60 **12.** 9.06
13. 26.32 **14.** 36.87 **15.** 49.04 **16.** 10 200
17. 161.8 **18.** 28 560 **19.** 4 207 000 **20.** 3.105×10^4
21. 11 **22.** 8 **23.** 7 **24.** 2.5
25. 1.721 **26.** 1.990 **27.** 6.293 **28.** 19.90
29. 62.93 **30.** 3.162 **31.** 10 **32.** 31.62
33. 100 **34.** 7.681 **35.** 24.52 **36.** 91.76
37. 2.881 **38.** 0.7308 **39.** 0.2530 **40.** 0.06
41. 2 **42.** 4 **43.** 0.178 **44.** 0.128
45. 0.010 41 **46.** 0.255 **47.** 0.039 **48.** 0.7508
49. 1.263 **50.** 1.577 **51.** 10.13 **52.** 0.094 67
53. 1.507 **54.** 0.5513 **55.** 0.030 67

b) **1.** 17.087 **2.** 31.52 **3.** 0.8463 **4.** 0.3967 **5.** 0.1241

CALCULATIONS 12

1. a) 1.7576 b) 1.7808 c) 1.0000 d) 3.0000 e) 3.9587 f) 0.8640
g) 2.6351 h) 0 i) $\bar{1}.0000$ j) $\bar{1}.4789$ k) $\bar{3}.6675$ l) $\bar{1}.9970$
m) $\bar{1}.5934$ n) $\bar{2}.7536$ o) $\bar{4}.5563$ p) $\bar{3}.8274$

2. 3, 0, 1, 1, −3, 1, −4, −2, 0, −4
3. −1, 2, 3, −3, −1, 5, 0, 4, −4, 2
4. $\bar{1}.7$, 2.8, $\bar{2}.8$, $\bar{1}.8$, 2.9, 2.5, 1.3, 0.0, 4.5, 0.2, 1.0, $\bar{1}.1$
5. $\bar{3}.1$, $\bar{4}.2$, 2.4, 5.2, 1.3, $\bar{1}.1$, $\bar{5}.9$, $\bar{3}.0$, $\bar{5}.1$, 1.2, 2.4, 4.1
6. a) 351.5 b) 479.1 c) 0.024 41 d) 0.048 46
e) 3.868 f) 33.52 g) 3.262 h) 9.092 i) 2.148 j) 0.060 46
k) 4.589 l) 7.701 m) 0.4432 n) 0.4067

CALCULATIONS 13

1. 374.055 54 2. 28.05 3. 0.023 92 4. 0.114 15
5. 828 6. 1.649×10^6 7. 2.2679 8. 0.019 76 9. 3198
10. 30.064 11. a) 2.708 b) 0.966 12. 8.336 13. a) 21.623
b) 2.780 14. 1.099×10^{-4}

SELF TEST NO. 1

1. a) $7\frac{1}{3}$ b) $2\frac{1}{3}$ c) $1\frac{11}{35}$ d) $\frac{7}{18}$ **2.** a) 8804 b) 154 c) 62.349
3. 108 kg 4. 4.2 days
5. a) 5 b) $7\frac{3}{4}$ c) $\frac{4}{5}$ **6.** a) 1.9375 b) $\frac{29}{32}$
7. 18.8% 8. 0.010 35
9. a) 1 1 1 0 1 1 b) 92 c) 1 1 0 1 0
10. a) 0.009 467 b) 19.34 c) 26.88 d) 17.21
11. 0.1411 **12.** 509.3

Section B

ALGEBRA

When you have completed this section you should be able to:

1. Represent quantities by numbers and letters.
2. Use the basic notation and rules of algebra.
3. Apply the precedence rules and laws of indices.
4. Multiply, divide and factorise expressions involving brackets.
5. Solve simple equations algebraically.
6. Solve simultaneous equations algebraically.
7. Derive and solve simple and simultaneous equations from practical problems.
8. Evaluate and manipulate formulae.

1. Find the value of A if:
   a) $A = 2x + x$ when $x = 3$
   b) $A = 3x + 4x$ when $x = 2$
   c) $A = 2x^2 + 3$ when $x = 4$
   d) $A = \dfrac{6}{x+2}$ when $x = 1$
   e) $A = \dfrac{x}{3+x^2}$ when $x = 4$
   f) $A = 2x + 3y$ when $x = 1$ $y = 2$
   g) $A = 4y \times 2x$ when $x = \frac{1}{2}$ $y = 2$
   h) $A = y^2 \times x^2$ when $x = 3$ $y = 4$
   i) $A = 2x + 3y + 4$ when $x = \frac{1}{2}$ $y = \frac{1}{3}$
   j) $A = \dfrac{y}{x+6}$ when $x = 1$ $y = 21$

2. If $p = 3$, $q = 1$, $s = 4, t = 0$, $r = 2$, find the value of:
   a) $4pq + 4rs - 2rt$
   b) $3prq - 2rst$
   c) $\dfrac{qr}{2} + \dfrac{rs}{4}$
   d) $\dfrac{pr - s}{q}$
   e) $\dfrac{2p + 2s}{3q + 2t}$
   f) $\dfrac{pq}{r + s}$
   g) $\dfrac{2s}{4r} - \dfrac{3q}{p} + \dfrac{2t}{1}$
   h) $\dfrac{3s}{p} \times \dfrac{q}{r}$
   i) $\dfrac{q + s}{3} \times \dfrac{1}{p + r}$
   j) $\dfrac{p}{s} \div \dfrac{q + r}{s - t}$

Write down the simplified forms of:

3. a) $3g + 4g + 5g$
   b) $3s + 4s - 2s$
   c) $8d - 2d + 3d$
   d) $6e - 3e - 2e$
   e) $k + 3k - 4k$
   f) $5y - 2y - 2y$
   g) $5x - 6x + 2x$
   h) $f + f - f$
   i) $e + e + e - 2e - 3e + 4e$
   j) $t - 4t - 2t + 3t + 4t$

4. a) $e + f + f + e$
   b) $p + 2q + 3q + 2q$
   c) $2a + 3a - 2a - 4a$
   d) $3n - 2p - 2n + 4p$
   e) $4k + 2 + k - p - 4$
   f) $2p + 3q - p - 2q - p - q$
   g) $6k - 3p - 2p$
   h) $3c - 4d - 2c + 3d - 1$
   i) $4 + 3g + 2h - 4$
   j) $4k - 2\ell - 3k + 4 + 2\ell - k$

5. a) $x \times x \times x$
   b) $2 \times a \times b \times 3 \times a$
   c) $2 \times x \times y$
   d) $5x \times 3y$
   e) $2ab \times 2ab$
   f) $x \times xy \times x$
   g) $xy \times xy \times 2$
   h) $2abc \times 3a$
   i) $3xyz \times 2xy$
   j) $2x \times 3y \times 0$

1. Remove the brackets and simplify:

a) $2(a+b)$
b) $3(a-3)$
c) $-2(x+y)$
d) $-3(x-y)$
e) $2a(a^3-a^2+a)$
f) $-3x^2(x^2+2xy+y^2)$
g) $(a+b)(a+b)$
h) $(a+b)(a-b)$
i) $(a+2)(a+3)$
j) $(a-3)(a-4)$
k) $(2a+b)(2a+2b)$
l) $(a+b)(x-y)$
m) $-(2a-b-c)$
n) $2(3s-4t)-5(s-2t)$
o) $3[5x-2(1+x)]$
p) $-3[6x-6(4x+7)-2(-x+4)]$
q) $(6t-7s)(9t+4s)$
r) $(2a+1)(a+2)(3a-3)$
s) $(a-1)(a+1)(a-1)$
t) $(a+b)^3$

2. Simplify:

a) $c \times c \times c$
b) $s \times s \times t$
c) $2ab \times 2ab$
d) $2x \times x \times y^2$
e) $3xyz \times 2xy$
f) $3c^4-2c^4-c^4$
g) $a^2b+2ab^2+a^2b$
h) $b^3c+bc^3+2b^3c$
i) $a+a^3-a^2+3a$
j) $x+x^3-2x^2+4x+x^2$
k) $m^5 \times m^2$
l) $x^2y \times y$
m) $mn \times nm$
n) $m^2n \times mn^2$
o) $abc^3 \times ac \times b^2c$
p) $a^2 \div a$
q) $m^3 \div m^2$
r) $b^3c^2 \div b^2c$
s) $p^2q^3r \div pqr$
t) $xy^2 \times yz^2 \div xyz$

3. a) Add (x^2-xy+y^2) to $(2x^2+xy-y^2)$
b) From (a^3-2a^2) subtract $(2a^2-3a+4)$
c) What must be added to $3x^2+4y^2$ to make $3x^2+2xy+4y^2$?
d) Expand $2a^2(3a+ab+1)$
e) Expand $-3x^2(x^2+2xy+y^2)$
f) Expand $(2a^2+3a)(3a^2-2a)$
g) Expand $(2a+1)^2$
h) Expand $(2x-2)(3x^3+2x^2-x+1)$
i) Expand $a(a^2+1)(a^2+2)$
j) Expand $(a-b)^3$

4. Expand:

a) $(2x)^3$

b) $(-2y)^3$

c) $(3a^2)^2$

d) $\sqrt{a^2}$

e) $\sqrt{4x^2}$

f) $\sqrt{\dfrac{a^4}{b^2}}$

g) $\sqrt{\dfrac{81}{9}}$

h) $\sqrt{\dfrac{4a^2}{9b^6}}$

i) $\sqrt[4]{81a^4}$

j) $\dfrac{\sqrt[3]{27x^3}}{3}$

5. Evaluate the following:

a) $64^{1/2}$

b) $27^{1/3}$

c) $(9^{1/2})^3$

d) $(36^{1/2})^3$

e) 10^{-1}

f) 5^{-2}

g) $64^{-2/3}$

h) $25^{1/2} \times 16^{1/2} \times 8^0$

i) $2^{3/2} \times 2^{-1/2} \times 64^{1/2}$

j) $\sqrt{3} \times 3^{1/2} \times 3^{-1}$

6. Rewrite in its more usual form:

a) $x^{1/2}$

b) x^{-1}

c) $x^{-1/3}$

d) x^{-2}

e) $x^{-2/3}$

f) $y^{3/2}$

g) $y^{-4/3}$

h) $6x^{-2}$

i) $8x^{-3/2}$

j) $9^{-1}x^{2/3}$

7. Rewrite the following in the above form:

a) $\dfrac{1}{x}$

b) $\dfrac{2}{x}$

c) $\sqrt{x}$

d) $\dfrac{4}{\sqrt[3]{x}}$

e) $\sqrt[3]{x^2}$

f) $\dfrac{3x}{\sqrt{x}}$

g) $\dfrac{1}{2x^2}$

h) $\dfrac{x^3}{\sqrt{y^2}}$

i) $\sqrt{x} \times \dfrac{1}{x}$

j) $\dfrac{1}{\sqrt{x}} \times x^2 \times \sqrt{x^3}$

1. Factorise the following:
   a) $ax + 3ay + 4az$
   b) $3t^2 - 9t^3$
   c) $mnx + mny$
   d) $ft - gt$
   e) $4t^4 + 32t^3 + 8t^2$
   f) $16ab^2 - 32a^2b$

2. Factorise the following:
   a) $x(4-z) + y(4-z)$
   b) $x(x+1) + 4(x+1)$
   c) $p(4m+n) - 4(4m+n)$
   d) $4p(x-6) - 3(-x+6)$
   e) $4m(m+3) + 3(-m-3)$
   f) $4t(x-y) + (x-y)$

3. Factorise the following by grouping in pairs:
   a) $ab - ac + eb - ec$
   b) $ut - vt - us + vs$
   c) $5xa + 25ya - 5xb - 25yb$
   d) $x^2 - 6x + 3x - 18$
   e) $t^2 - 7t - 4t + 28$
   f) $x^2 - ax - x + a$
   g) $ep - fp + eq - fq$
   h) $ak + ap - bk - bp$
   i) $6xy + 9x + 4y + 6$
   j) $xy - 4x - 3y + 12$
   k) $6ac + 3bc - 2ad - bd$
   l) $a^2x^2 + abx - adx - bd$

4. Factorise the following:
   a) $x^2 + 10x + 16$
   b) $x^2 + 11x + 30$
   c) $x^2 - 8x + 12$
   d) $x^2 + 5x - 14$
   e) $t^2 + 15t - 16$
   f) $a^2 - 5a - 36$
   g) $a^2 - 7a + 10$
   h) $y^2 - 9y + 20$
   i) $y^2 - 8y - 65$
   j) $x^2 - 7x - 18$
   k) $8 - 9x + x^2$
   l) $117 - 4x - x^2$

5. Factorise the following differences:
   a) $p^2 - q^2$
   b) $25t^2 - 9r^2$
   c) $m^2 - 16n^2$
   d) $45y^2 - 20z^2$
   e) $300p^2 - 12j^2$
   f) $4x^2 - 9y^2$

6. Simplify the following:
   a) $\dfrac{x^2 - 4x + 3}{x - 3}$
   b) $\dfrac{x^2 + 6x + 9}{x + 3}$
   c) $\dfrac{x^2 - 7x + 12}{x - 4}$
   d) $\dfrac{x^2 + x - 30}{x + 6}$

1. Multiply out or expand the following:

a) $(x+2)(x+1)$
b) $(y-3)(y+4)$
c) $(2z-6)(z-2)$
d) $(-x+5)(3x+7)$
e) $(x+y)(x-2y)$
f) $(a-b)(a+5b)$
g) $(8+y)(8+y)$
h) $(x+7)^2$
i) $(y+2u)^2$
j) $(x-y)(x+y)$
k) $(v-4u)(v+4u)$
l) $(6t-7s)(9t+4s)$

2. Simplify the following:

a) $\frac{2x}{5}+\frac{3x}{4}$
b) $\frac{2t}{3}-\frac{5t}{7}+\frac{9t}{21}$
c) $\frac{(x-3)}{4}+\frac{(x-2)}{7}$
d) $\frac{(2s-7)}{9}-\frac{(5s+2)}{6}$
e) $\frac{4(2z+5)}{5}-\frac{2(4-2z)}{7}$
f) $\frac{(x-3)}{4}+\frac{7}{(x-2)}$
g) $\frac{3t}{4}-\frac{(5t-4)}{3}$
h) $5-\frac{(2x+4)}{3}$
i) $\frac{(y-3)(y-4)}{12}$
j) $\frac{(a+b)(a-b)}{a^2}$

3. Express the following in their simplest form:

a) $\frac{a}{x}+\frac{a}{3x}-\frac{2a}{4x}$
b) $\frac{m}{nq}+\frac{n}{mq}+\frac{q}{mn}$
c) $\frac{1}{x+1}+\frac{1}{x-1}$
d) $\frac{a}{a+b}-\frac{a}{a-b}$
e) $\frac{x+3}{x-3}-\frac{x-3}{x+3}$
f) $\frac{3}{1-a}+\frac{4}{(1-a)^2}$
g) $\frac{a-b}{c-d}-\frac{b-a}{d-c}$
h) $\frac{2b}{a-2b}+\frac{a}{a+2b}$
i) $\frac{1}{p+q}+\frac{1}{p-q}+\frac{2p}{p^2-q^2}$
j) $\frac{1}{a^2-5a+4}-\frac{1}{a^2-4a+3}$

1. Find x when $6x = 4.5x + 18$
2. Solve $7x + 10 = 4x + 19$
3. Solve $5(3x - 4) = 40$
4. Solve $3x + 5 = x + (3x - 12)$
5. For what value of n is $3n + 7$ equal to $14 + 2.5n$?
6. Solve $12r - 5(r - 1) = 2r + 6$
7. Solve: $4(x + 2) - 3(4 - x) + 24 = 34$
8. Solve: $5(x + 2) - 3(x - 3) = 23$
9. Solve: $3(x - 1) - 4(2 - 3x) = 19$
10. Solve $(1 - x) - 3(x - 4) = 33$
11. Solve $2t - 4 = 3(t - 1.6)$
12. Solve $2n = 0.58\,(12 - n)$
13. Solve $3(n - 7) = 6 - 4(3 - n)$
14. For what value of r is 18.4 equal to $2(3.5r - 1)$?
15. Find n when $15.8 = \dfrac{56}{3n}$
16. Find n when $\dfrac{7.5}{n} = \dfrac{5}{2}$
17. Find c if $\dfrac{18}{2c} = 3.8$
18. If $c = \dfrac{V}{R}$
    a) Find V when $c = 8, R = 4.5$
    b) Find R when $c = 7.5, V = 60$
19. For what value of x is $3(x - 5)$ equal to $\dfrac{4x + 3}{2}$?

Solve the following equations:

20. $\dfrac{x}{3} - \dfrac{x}{4} = \dfrac{1 - x}{6}$
21. $\dfrac{3 - x}{4} = \dfrac{x}{6}$
22. $n - 7 = \dfrac{2n - 5}{6}$
23. $\dfrac{x - 1}{x - 2} = 3$
24. $\dfrac{1 - r}{r + 1} = 4$
25. $2 = \dfrac{100 - t}{360 - t}$
26. $0.8 = \dfrac{1.5n}{4 + 1.4n}$
27. $\frac{1}{5}(x + 3) - \frac{1}{2}(x + 2) = \frac{1}{4}(x + 8)$
28. $\dfrac{3p + 6}{5} - \dfrac{4p + 3}{4} = \dfrac{2p + 7}{6}$
29. $\dfrac{3p + 23}{3p + 12} = \dfrac{4}{3}$
30. $(R - 3)(2R + 6) = 2R(R - 18)$
31. $1 + 0.0062x = \dfrac{67.6 \times 1.03}{25.6}$
32. $\dfrac{5}{2x + 5} = \dfrac{4}{x + 5}$

PROBLEMS LEADING TO SIMPLE EQUATIONS (UNGRADED)

1. Find a number such that when 8 is added to it, the result is twice as much as when 4 is subtracted from it.

2. In an isosceles triangle the base is two-thirds of one of the equal sides and the sum of the sides is 40 mm.
   Find the three sides.

3. Find the angles of an isosceles triangle in which each of the angles at the base is double the vertical angle.

4. When the two pieces of a broken metre-stick are placed side by side one of them is seen to be $4\frac{1}{2}$ cm longer than the other.
   At what point did the stick break?

5. Two bags of bolts are of equal mass, but there are 50 bolts more in one bag than the other. The bolts in one bag have a mass of 50 g each, in the other 45 g each.
   How many bolts are there in each bag?

6. A wire was measured in inches and also in centimetres.
   If the number of centimetres was greater than the number of inches by 231 what was the length in inches (1 in = 2.54 cm)?

7. A man took a handful of nails for a job, twice as many as he needed. Eight of the nails were the wrong size, but three quarters of those that were left finished the job.
   How many did the job require?

8. Wood's fusible alloy (melting point 61 °C) contains equal quantities of tin and cadmium, twice as much lead as tin, and twice as much bismuth as lead.
   What mass of cadmium is contained in 1 kg of the alloy?

9. An employee produces a total of 910 articles. He commences at the rate of 20 articles per hour and, having completed a part of the total, increases his rate to 25 articles per hour. The time taken overall was 40 hours.
   How many articles had been made when his production rate was changed?

10. What mass of tin must be melted with 48 kg of copper in order to produce a bronze containing 16.5% tin?

11. What weight of zinc sulphate must be added to 45 grams of water in order to prepare a 35% solution of the sulphate in water?

Solve the following simultaneous equations:

1. $3x - 2y = 7$
$x + 2y = 5$

2. $2x + 3y = 5$
$x + y = 2$

3. $5x - 2y = 5$
$6y - 2x - 11$

4. $2P - 5Q = 2$
$3P + 10Q = 8.6$

5. $5x + 2y = 16$
$x + 3y = 11$

6. $5x - y = 14$
$4x - 2y = 4$

7. $3x - 2y = 7$
$2x + 3y = 22$

8. $2x - 3y = 13$
$5x + 2y = 4$

9. $8x - 3y = 39$
$7x + 5y = -4$

10. $0.1x + 0.2y = -0.2$
$1.5x - 0.4y = 10.6$

11. $\frac{1}{2}x + \frac{1}{3}y = 3$
$\frac{7}{8}x + \frac{1}{2}y = 2$

12. $3x - 4y = 10$
$\frac{x}{5} - \frac{y}{4} = 1$

13. $x + \frac{4}{y} = 11$
$2x - \frac{1}{y} = 4$

14. $\frac{x}{3} + \frac{y}{4} = 6$
$\frac{x}{6} - \frac{y}{8} = 0$

15. $\frac{1}{P} + \frac{1}{Q} = \frac{7}{12}$
$\frac{1}{P} - \frac{1}{Q} = \frac{1}{12}$

1. A model made of 20 cm^2 of copper plate and 15 cm^2 of aluminium plate has a mass of 215 g. If 25 cm^2 of copper and 10 cm^2 of aluminium were used, it would have a mass of 246 g.
   Find the mass per sq cm of each kind of plate.

2. A new house requires lamps for 16 lighting 'points'. If 4 of 100 W and 12 of 60 W are fitted the cost for lamps is £3. If 6 are 100 W and 10 are 60 W the cost is £3.16.
   Find the price of each kind of lamp.

3. A parliamentary candidate addressed the electors on two nights. The average number present per night was 278. 10 enthusiasts attended on both nights. If they had not attended on the second night the attendance would have been only half that on the first night.
   How many were present each night?

4. Two gear wheels mesh with one another, and while one makes 3 turns the other makes 10. If each of them were cut with 12 teeth more, the first would make 3 turns while the other made 8.
   How many teeth are in each wheel?

5. Find two numbers whose sum is 100 and whose difference is 44.

6. Two quantities, X and Y, are connected by a law of the form $Y = mX + b$. When $X = 1$, $Y = 32$, and when $X = 10$, $Y = 95$.
   Find the connection between X and Y.

7. In testing a crane it was found that a load of 1000 N was raised by an effort of 77.4 N, and a load of 500 N by an effort of 41.7 N.
   Find a formula connecting the effort (E) with the load (W), assuming that it is of the type $E = aW + b$.

8. From London to Crawley is 52 km by road. Alex leaves London for Crawley at 8.00 a.m. If Bill leaves Crawley for London at 8.50 a.m. they meet at 10.50 a.m. If however Bill did not leave until 10 a.m. they would have met at 11.20 a.m.
   At what speed does each travel?

Derive a *formula* from each of the following statements:

1. The area of a triangle is formed by multiplying the length of the base by the perpendicular height and dividing the result by 2.

2. The power consumed by an electric lamp is found by multiplying the voltage used by the current it takes.

3. The velocity ratio (R) of a machine is found by dividing the distance moved by the load (l) into the distance moved by the effort (e).

4. The volume of a sphere is the product of π (pi) and the cube of its radius all multiplied by 4/3.

5. a is equal to 5 plus the product of b and c.

Derive a formula for:

6. The perimeter (p) of a triangle with sides a, $(a + 2)$ and $(a - 2)$.

7. The perimeter of a square of side length $(2x - 3)$.

8. The base (b) of an isosceles triangle whose equal sides are $(a - 1)$ and perimeter P (an isosceles triangle has two sides of equal length).

Translate the following into statements similar to **5**:

9. $V = u + ft$.

10. $V = \pi r^2 h$.

Construct *formulae* to express the following:

11. The perimeter of a triangle (P) in terms of the length of its sides (a) (b) and (c).

12. The area of a triangle, which is one quarter the area of a rectangle measuring a by b.

13. The difference between two numbers is d and the larger number is n. Give their product (P) the average (A).

14. Two cars travel in opposite directions from the same town. One at x km/hr and the other at twice that speed.
    How far apart are they after T hours?

Derive a formula for the following:

15. A spring is a mm long and extends b mm for each weight supported. What is its total length when supporting w weights?

16. A man's ordinary rate of pay is a pence per hour but b pence for overtime.
What will he receive after working 12 hours including 2 hours overtime? (in pounds).

17. A box when empty weighs x kN and when full of bolts weighs y kN. What will be the total weight when half full of bolts?

18. A cylindrical can D mm diameter is L mm long.
What is its total surface area including top and bottom?

19. Find the volume of a ring formed from rectangular section bar whose outside diameter is D, inside diameter is d and thickness is T.

20. Find the cost of n rectangular sheets of metal each measuring a m by b m at d pence per m^2.

21. Find the total wages paid to n men in one week when the rate of pay is p pence per hour for a 40 hour week.

22. A wheel has n spokes.
What is the angle in degrees between two adjacent spokes?

23. A rectangular plate length a and breadth b is cut from one measuring c by d.
What is the area of plate left?

24. The floor of a factory is b long by c wide. It is divided into two so that one part is twice the area of the other.
What is the area of the smallest part?

25. A lathe cost £M and has since depreciated in value by N per cent.
What is its present value?

26. The rent of a workshop is £R per annum, and repairs to it average one-twentieth of the rent.
What is the total outlay over a period of N years?

Give a formula for **a)** the area and **b)** the perimeter of:

27. **28.** **29.** **30.**

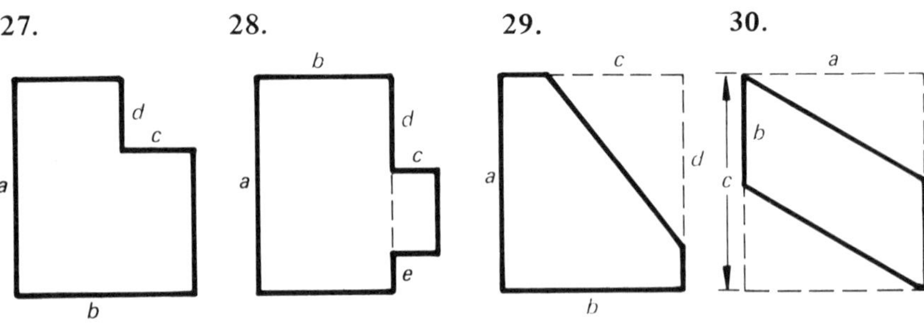

1. Change the subject of the following formulae to the letter indicated:

a) $x = 3 + s$ (s)
b) $g = p + 7q + 9x$ (p)
c) $s = m \cdot n$ (n)
d) $p = \frac{q}{s}$ (q)
e) $a = \frac{5b}{c}$ (c)
f) $y = x - c$ (c)
g) $p = 4y - 2$ (y)
h) $k = wp + 2q$ (p)
i) $a = b(c + 2d)$ (b)
j) $z = 2(c + d)$ (d)

2. Change the subject of the following formulae to the letter indicated:

a) $v = u + at$ (t)
b) $p = \frac{Rt}{V}$ (V)
c) $E = C(R + r)$ (R)
d) $E = \frac{1}{2}mv^2$ (m)
e) $s = ut + \frac{1}{2}at^2$ (u)
f) $R = \frac{S + C}{C}$ (S)
g) $l = l_0(1 + \alpha t)$ (α)
h) $V = \frac{\pi d^2 h}{4}$ (h)
i) $H = \frac{d^2 s}{12}$ (s)
j) $H = (W + w)(T - t)$ (t)

3. Evaluate the following formulae taking π as $\frac{22}{7}$:

a) $V = u + ft$ $u = 16$ $f = 32$ $t = 2\frac{1}{2}$
b) $s = ut + \frac{1}{2}ft^2$ $u = 16$ $f = 32$ $t = 2$
c) $v^2 = u^2 + 2fs$ $u = 12$ $f = 32$ $s = 4$ (find v)
d) $\frac{1}{f} = \frac{1}{V} - \frac{1}{u}$ $u = 10$ $V = 8$ (find f)
e) $I = \frac{M}{3}(a^2 + b^2)$ $M = 51$ $a = 8$ $b = 4$
f) $V = \frac{4\pi r^3}{3}$ $r = 3$
g) $A = 2\pi r(h + r)$ $r = 2$ $h = 1\frac{1}{2}$
h) $t = \frac{2\pi A}{Cn}$ $A = 4$ $C = 3$ $n = 55$
i) $p = \frac{2\pi M'}{M + M'}$ $M' = 4$ $M = 3$
j) $T = 2\pi\sqrt{\frac{l}{g}}$ $l = 128$ $g = 32$

1. Re-arrange the formula to make the subject the letter indicated: (Take care as the subject may appear more than once.)

a) $E = \dfrac{(a-b)W}{2aP}$ (a)

b) $T = \dfrac{M}{r}(u^2 - 5gr)$ (r)

c) $n = \dfrac{2a}{a-b}$ (a)

d) $\dfrac{I}{M} = \dfrac{r-c}{r(c-1)}$ (c)

e) $\dfrac{I}{M} = \dfrac{r-c}{r(c-1)}$ (r)

f) $T = t + (At - K)$ (t)

g) $S = \dfrac{W}{W-w}$ (W)

h) $C = 2\sqrt{2rh - h}$ (h)

i) $\dfrac{W}{w} = \dfrac{D^2}{d^2}$ (d)

j) $T = 2\pi\sqrt{\dfrac{I}{MH}}$ (M)

k) $\dfrac{1}{V} = \dfrac{1}{u} + \dfrac{1}{f}$ (V)

l) $H = \left(\dfrac{1+at}{1+bt}\right)h$ (t)

m) $R = \dfrac{s+c}{c}$ (c)

n) $C = \dfrac{ME}{R+Mr}$ (M)

o) $M = \dfrac{2a}{a-b}$ (b)

2. Re-arrange the formula and evaluate to find the letter indicated:

	Formula	Values	Find
a)	$A = LB$	$A = 12,\ B = 3$	(L)
b)	$V = L \cdot B \cdot H$	$V = 48,\ L = 6,\ H = 4$	(B)
c)	$P = \dfrac{1}{n}$	$P = 12$	(n)
d)	$A = \dfrac{V}{L}$	$A = 15,\ V = 125$	(L)
e)	$L = D + 2K$	$L = 5,\ D = 4.5$	(K)
f)	$d = 2C - D$	$d = 1.41,\ D = 4.09$	(C)
g)	$R = R_1 + R_2 + R_3$	$R = 14,\ R_1 = 4,\ R_2 = 3$	(R_3)
h)	$V = u + gt$	$V = 815,\ u = 15,\ g = 32$	(t)
i)	$V = \pi R^2 h$	$V = 5\frac{1}{7},\ \pi = 3\frac{1}{7},\ h = 6\frac{6}{11}$	(R)
j)	$V = \dfrac{\pi d^2 h}{4}$	$V = 115\frac{1}{2},\ \pi = 3\frac{1}{7},\ h = 12$	(d)
k)	$P = \dfrac{N + n}{2c}$	$P = 6,\ c = 7,\ N = 60$	(n)
l)	$C = \dfrac{N - n}{2P}$	$N = 100,\ C = 5,\ P = 6$	(n)
m)	$P = \dfrac{S(c - F)}{c}$	$P = \frac{1}{5},\ S = \frac{1}{4},\ F = \frac{4}{5}$	(c)
n)	$H = 0.06C^2RT$	$H = 40\,000,\ C = 10,\ R = 600$	(T)
o)	$s = \pi r(r + h)$	$s = 99,\ \pi = 3\frac{1}{7},\ r = 3\frac{1}{2}$	(h)
p)	$R = r + art$	$R = 230,\ a = 0.005,\ t = 30$	(r)
q)	$I = \dfrac{E}{R + r}$	$E = 2.5,\ I = 0.05,\ R = 30$	(r)
r)	$A = \dfrac{\pi}{4}(D^2 - d^2)$	$A = 2\frac{3}{4},\ D = 3\frac{3}{4},\ \pi = 3\frac{1}{7}$	(d)
s)	$r = \sqrt{(0.36A}$	$r = 4.2$	(A)
t)	$a = \sqrt{b^2 + c^2}$	$a = 5,\ c = 3$	(b)
u)	$a = b + \sqrt{b^2 + c^2}$	$a = 6,\ c = 4$	(b)
v)	$T = 2\pi\sqrt{\dfrac{l}{g}}$	$T = 2\frac{5}{14},\ l = 4\frac{1}{2},\ \pi = 3\frac{1}{7}$	(g)
w)	$Ah = 12(P - Q)$	$A = \frac{3}{5},\ h = 20,\ Q = 3$	(P)
x)	$l = ar^2$	$l = 36,\ a = 4$	(r)

Transpose the following formulae and evaluate:

1. $x^2 + y^2 = r^2$	$y = 8,\ r = 10$	(find x)
2. $\dfrac{x^2}{a^2} + \dfrac{y^2}{b^2} = 1$	$y = 3,\ b = 5,\ a = 6$	(find x)
3. $l = l_0(1 + \alpha t)$	$l = 122.4,\ l_0 = 120,\ t = 50$	(find α)
4. $p = \dfrac{k}{v}$	$p = 800,\ k = 96\,000$	(find v)
5. $I = \dfrac{M}{4}(a^2 + b^2)$	$I = 113\frac{3}{4},\ a = 3\frac{1}{2},\ b = 2$	(find M)
6. $H = \dfrac{2\pi IT}{l}$	$H = 8000,\ I = 10,\ l = 5\frac{1}{2}$	(find T)
7. $\sigma = \dfrac{\rho W}{W - W'}$	$\sigma = 2.4,\ W = 12,\ W' = 8$	(find ρ)
8. $p = aW + b$	$p = 100,\ W = 2240,\ b = 20$	(find a)
9. $D = d(1 + at)$	$D = 50.096,\ d = 50,\ t = 150$	(find a)
10. $n = \dfrac{K\sqrt{T}}{L}$	$n = 128,\ T = 4500,\ L = 61.8$	(find K)
11. $\dfrac{1}{R} = \dfrac{1}{R_1} + \dfrac{1}{R_2} + \dfrac{1}{R_3}$	$R = 2.5,\ R_1 = 7,\ R_3 = 9$	(find R_2)
12. $\dfrac{T - t}{t} = \dfrac{\mu\theta - 1}{\mu\theta}$	$T = 21,\ t = 20,\ \theta = 3$	(find μ)

1. Find the value of A when:
   a) $A = 5 - 2x$ and $x = \frac{1}{2}$
   b) $A = 4y - 3x$ and $x = -2, y = \frac{1}{4}$

2. Simplify:
   a) $5x + 5t - 4t$
   b) $3ab \times 4ab$
   c) $(3a - 1)(a + 2)$
   d) $(2x + y)^2$
   e) $(2x)^4$
   f) $\sqrt[3]{27a^6}$
   g) $(x^2 + 2xy) - (x^2 - 3xy)$
   h) $b^2c^3 \div b^2c$

3. Factorise:
   a) $3x^3 + 6x^2 - 9x$
   b) $x^2 - 4x + 3$
   c) $4t^2 - 25s^2$
   d) $ab - bc - cd + da$

4. Expand:
   a) $(2x + y)^2$
   b) $(3a + 4)(2a - 6)$

5. Simplify:
   a) $\dfrac{(x+2)}{3} - \dfrac{(2x-3)}{4}$
   b) $\dfrac{1}{x-1} + \dfrac{1}{x+1}$

6. Solve to find x:
   a) $6x = 2x + 12$
   b) $4(x + 3) - 2(x - 7) = 30$
   c) $\dfrac{3.5}{2x} = \dfrac{3}{4}$
   d) $\dfrac{3-x}{3} = \dfrac{x}{5}$

7. Find the value which when added to 9 gives three times as many as when added to 2.

8. Solve the simultaneous equation $2x + 3y = -2$, $3x + 2y = +2$.

9. Two quantities x and y are connected by a law of the form $y = Mx + C$.
   Find M and C if when $x = 4$, $y = 5$ and when $x = 12$, $y = 9$.

10. A box when empty weighs x (N) and when full weighs y (N). Construct a formula for the weight of two boxes each half full.

11. Change the subject of the formula to the letter indicated:
    a) $K = Wp + 2q$ (W)
    b) $l = l_0(1 + \alpha t)$ (t)
    c) $V = \dfrac{\pi d^2 h}{4}$ (d)
    d) $a = \dfrac{2b}{b-c}$ (b)

12. Evaluate the following formulae:
    a) $V = u + ft$ $\quad u = 3, \quad f = 32, \quad t = 1\frac{1}{2}$
    b) $V = \dfrac{\pi d^2 h}{4}$ $\quad \pi = 3\frac{1}{7}, \quad d = 3, \quad h = 14$
    c) $P = \dfrac{s(c-F)}{c}$ $\quad S = 13, \quad c = 5, \quad F = 3$
    d) $\dfrac{1}{R} = \dfrac{1}{R_1} + \dfrac{1}{R_2} + \dfrac{1}{R_3}$ $\quad R_1 = 4, \quad R_2 = 5, \quad R_3 = 6$

ALGEBRA 1

1. a) 9 b) 14 c) 35 d) 2 e) $\frac{4}{19}$
   f) 8 g) 8 h) 144 i) 6 j) 3
2. a) 44 b) 18 c) 3 d) 2 e) $4\frac{2}{3}$
   f) $\frac{1}{2}$ g) 0 h) 2 i) $\frac{1}{3}$ j) 1
3. a) $12g$ b) $5s$ c) $9d$ d) e e) 0
   f) y g) x h) f i) $2e$ j) $2t$
4. a) $2e + 2f$ b) $p + 7q$ c) $-a$ d) $n + 2p$ e) $5k - p - 2$
   f) 0 g) $6k - 5p$ h) $c - d - 1$ i) $3g + 2h$ j) 4
5. a) x^3 b) $6a^2b$ c) $2xy$ d) $15xy$ e) $4a^2b^2$
   f) x^3y g) $2x^2y^2$ h) $6a^2bc$ i) $6x^2y^2z$ j) 0

ALGEBRA 2

1. a) $2a + 2b$ b) $3a - 9$ c) $-2y - 2x$
   d) $3y - 3x$ e) $2a^4 - 2a^3 + 2a^2$ f) $-3x^4 - 6x^3y - 3x^2y^2$
   g) $a^2 + 2ab + b^2$ h) $a^2 - b^2$ i) $a^2 + 5a + 6$
   j) $a^2 - 7a + 12$ k) $4a^2 + 6ab + 2b^2$ l) $ax + bx - ay - by$
   m) $b + c - 2a$ n) $s + 2t$ o) $9x - 6$
   p) $48x + 150$ q) $54t^2 - 39st - 28s^2$ r) $6a^3 + 9a^2 - 9a - 6$
   s) $a^3 - a^2 - a + 1$ t) $a^3 + 3a^2b + 3ab^2 + b^3$
2. a) c^3 b) s^2t c) $4a^2b^2$
   d) $2x^2y^2$ e) $6x^2y^2z$ f) 0
   g) $2a^2b + 2ab^2$ h) $3b^3c + bc^3$ i) $4a - a^2 + a^3$
   j) $x^3 - x^2 + 5x$ k) m^7 l) x^2y^2
   m) m^2n^2 n) m^3n^3 o) $a^2b^3c^5$
   p) a q) m r) bc
   s) pq^2 t) y^2z
3. a) $3x^2$ b) $a^3 - 4a^2 + 3a - 4$ c) $2xy$
   d) $6a^3 + 2a^3b + 2a^2$ e) $-3x^4 - 6x^3y - 3x^2y^2$ f) $6a^4 + 5a^3 - 6a^2$
   g) $4a^2 + 4a + 1$ h) $6x^4 - 2x^3 - 6x^2 + 4x - 2$ i) $a^5 + 3a^3 + 2a$
   j) $a^3 - 3a^2b + 3ab^2 - b^3$
4. a) $8x^3$ b) $-8y^3$ c) $9a^4$ d) a e) $2x$
   f) $\dfrac{a^2}{b}$ g) 3 h) $\dfrac{2a}{3b^3}$ i) $3a$ j) x
5. a) 8 b) 3 c) 27 d) 216 e) $\frac{1}{10}$
   f) $\frac{1}{25}$ g) $\frac{1}{16}$ h) 20 i) 16 j) 1
6. a) $\sqrt{x}$ b) $\dfrac{1}{x}$ c) $\dfrac{1}{\sqrt[3]{x}}$ d) $\dfrac{1}{x^2}$ e) $\dfrac{1}{\sqrt[3]{x^2}}$
   f) $\sqrt{y^3}$ g) $\dfrac{1}{\sqrt[3]{y^4}}$ h) $\dfrac{6}{x^2}$ i) $\dfrac{8}{\sqrt{x^3}}$ j) $\dfrac{\sqrt[3]{x^2}}{9}$
7. a) x^{-1} b) $2x^{-1}$ c) $x^{1/2}$ d) $4x^{-1/3}$ e) $x^{2/3}$
   f) $3x^{1/2}$ g) $\frac{1}{2}x^{-2}$ h) x^3y^{-1} i) $x^{-1/2}$ j) x^3

ALGEBRA 3

1. a) $a(x + 3y + 4z)$ b) $3t^2(1 - 3t)$ c) $mn(x + y)$
   d) $t(f - g)$ e) $4t^2(t^2 + 8t + 2)$ f) $16ab(b - 2a)$
2. a) $(4 - z)(x + y)$ b) $(x + 1)(x + 4)$ c) $(4m + n)(p - 4)$
   d) $(4p + 3)(x - 6)$ e) $(4m - 3)(m + 3)$ f) $(x - y)(4t + 1)$
3. a) $(b - c)(a + e)$ b) $(u - v)(t - s)$ c) $5(a - b)(x + 5y)$
   d) $(x + 3)(x - 6)$ e) $(t - 4)(t - 7)$ f) $(x - a)(x - 1)$
   g) $(p + q)(e - f)$ h) $(a - b)(k + p)$ i) $(3x + 2)(2y + 3)$
   j) $(x - 3)(y - 4)$ k) $(2a + b)(3c - d)$ l) $(ax + b)(ax - d)$
4. a) $(x + 2)(x + 8)$ b) $(x + 5)(x + 6)$ c) $(x - 2)(x - 6)$
   d) $(x + 7)(x - 2)$ e) $(t + 16)(t - 1)$ f) $(a + 4)(a - 9)$
   g) $(a - 5)(a - 2)$ h) $(y - 5)(y - 4)$ i) $(y + 5)(y - 13)$

j) $(x-9)(x+2)$ k) $(x-8)(x-1)$ l) $(13+x)(9-x)$

5. a) $(p+q)(p-q)$ b) $(5t-3r)(5t+3r)$ c) $(m+4n)(m-4n)$
d) $5(3y+2z)(3y-2z)$ e) $12(5p+j)(5p-j)$ f) $(2x+3y)(2x-3y)$

6. a) $(x-1)$ b) $(x+3)$ c) $(x-3)$ d) $(x-5)$

ALGEBRA 4

1. a) x^2+3x+2 b) y^2+y-12 c) $2z^2-10z+12$
d) $-3x^2+8x+35$ e) $x^2-xy-2y^2$ f) $a^2+4ab-5b^2$
g) $64+16y+y^2$ h) $x^2+14x+49$ i) $y^2+4yu+4u^2$
j) x^2-y^2 k) v^2-16u^2 l) $54t^2-39st-28s^2$

2. a) $\frac{23x}{20}$ b) $\frac{8t}{21}$ c) $\frac{11x-29}{28}$
d) $\frac{-11s-20}{18}$ e) $\frac{76z+100}{35}$ f) $\frac{x^2-5x+34}{4(x-2)}$
g) $\frac{16-11t}{12}$ h) $\frac{11-2x}{3}$ i) $\frac{y^2-7y+12}{12}$
j) $\frac{a^2-b^2}{a^2}$

3. a) $\frac{5a}{6x}$ b) $\frac{m^2+n^2+q^2}{mnq}$ c) $\frac{2x}{x^2-1}$
d) $\frac{2ab}{b^2-a^2}$ e) $\frac{12x}{x^2-9}$ f) $\frac{7-3x}{(1-a)^2}$
g) 0 h) $\frac{a^2+4b^2}{a^2-4b^2}$ i) $\frac{4p}{p^2-q^2}$
j) $\frac{1}{(a-1)(a-3)(a-4)}$

ALGEBRA 5

1. 12	2. 3	3. 4	4. 17	5. 14
6. $\frac{1}{5}$	7. 2	8. 2	9. 2	10. -5
11. 0.8	12. 2.7	13. -15	14. 2.914	15. 1.181
16. 3	17. 2.368	18. 36/8	19. 16.5	20. $\frac{2}{3}$
21. $\frac{9}{5}$	22. $9\frac{1}{4}$	23. $2\frac{1}{2}$	24. $-\frac{3}{5}$	25. 620
26. 8.42	27. -4.36	28. -0.977	29. 7	30. $\frac{1}{2}$
31. 277.4	32. $1\frac{2}{3}$			

ALGEBRA 6

1. 16	2. 15/10	3. 36°/72°	4. 47.75/52.25	5. 450/500
6. 150	7. 12	8. 125g	9. 360	10. 9.485 kg
11. 24.23g				

ALGEBRA 7

1. $x=3, y=1$	2. $x=1, y=1$	3. $x=2, y=\frac{5}{2}$
4. $P=1.8, Q=0.32$	5. $x=2, y=3$	6. $x=4, y=6$
7. $x=5, y=4$	8. $x=2, y=-3$	9. $x=3, y=-5$
10. $x=6, y=-4$	11. $x=-20, y=39$	12. $x=30, y=20$
13. $x=3, y=\frac{1}{2}$	14. $x=9, y=12$	15. $P=3, Q=4$

ALGEBRA 8

1. Cu 8.8, Al 2.6 2. 100W = 24.75 p, 60 W = 16.75 p 3. 364, 192
4. 100, 30 5. 72, 28 6. $y=7x+25$
7. $E=0.0714\,W+6$ 8. 12 km/hr, 9 km/hr

ALGEBRA 9

No answers

ALGEBRA 10

1. a) $x-3$ b) $g-7q-9x$ c) $\frac{s}{m}$ d) ps e) $\frac{5b}{a}$
   f) $x-y$ g) $\frac{P+2}{4}$ h) $\frac{k-2q}{w}$ i) $\frac{a}{c+2d}$ j) $\frac{z}{2}-c$
2. a) $\frac{v-u}{a}$ b) $\frac{Rt}{p}$ c) $\frac{E}{C}-r$ d) $\frac{2E}{v^2}$ e) $\frac{s}{t}-\frac{1}{2}at$
   f) $C(R-1)$ g) $\frac{l-l_0}{l_0 t\alpha}$ h) $\frac{4V}{\pi d^2}$ i) $\frac{12H}{d^2}$ j) $T-\frac{H}{(W+w)}$
3. a) 96 b) 96 c) 20 d) 40 e) 1360
   f) $113\frac{1}{7}$ g) 44 h) 0.152 i) 3.591 j) $12\frac{4}{7}$

ALGEBRA 11

1. a) $\frac{bW}{W-2EP}$ b) $\frac{Mu^2}{T+5Mg}$ c) $\frac{nb}{n-2}$
   d) $\frac{r(M+I)}{Ir+M}$ e) $\frac{Mc}{M+I-Ic}$ f) $\frac{T+K}{1+A}$
   g) $\frac{Sw}{S-1}$ h) $\frac{C^2}{4(2r-1)}$ i) $\sqrt{\frac{wD^2}{W}}$
   j) $\frac{4\pi^2 I}{T^2 H}$ k) $\frac{fu}{f+u}$ l) $\frac{h-H}{Hb-ha}$
   m) $\frac{S}{R-1}$ n) $\frac{CR}{E-Cr}$ o) $a-\frac{2a}{M}$
2. a) 4 b) 2 c) $\frac{1}{12}$ d) $8\frac{1}{3}$ e) 0.25
   f) 2.75 g) 7 h) 25 i) $\frac{1}{2}$ j) $3\frac{1}{2}$
   k) 24 l) 40 m) 4 n) 11.11 o) $5\frac{1}{2}$
   p) 200 q) 20 r) $3\frac{1}{4}$ s) 49 t) 4
   u) $1\frac{2}{3}$ v) 32 w) 4 x) 3

ALGEBRA 12

1. 6 2. $4\frac{4}{5}$ 3. 0.0004 4. 120 5. 28
6. 700 7. 0.8 8. $\frac{1}{28}$ 9. 1.28×10^{-5} 10. 118
11. 6.848 12. 0.35

SELF TEST NO. 2

1. a) 4, b) 7
2. a) $5x+t$ b) $12a^2b^2$ c) $3a^2+5a-2$ d) $4x^2+4xy+y^2$
   e) $16x^4$ f) $3a^2$ g) $5xy$ h) c^2
3. a) $3x(x-1)(x+3)$ b) $(x-1)(x-3)$ c) $(2t+5s)(2t-5s)$
   d) $(a-c)(b+d)$
4. a) $4x^2+4xy+y^2$ b) $6a^2-10a-24$
5. a) $\frac{17-2x}{12}$ b) $\frac{2x}{x^2-1}$
6. a) 3 b) 2 c) $2\frac{1}{3}$ d) $1\frac{7}{8}$
7. $x=1\frac{1}{2}$ 8. $y=-2, x=2$ 9. $M=\frac{1}{2}, c=3$ 10. $y+x$
11. a) $W=\frac{K-2q}{p}$ b) $t=\frac{l-l_0}{l_0\alpha}$ c) $d=\sqrt{\frac{4V}{\pi h}}$ d) $b=\frac{ac}{a-2}$
12. a) $V=51$ b) $V=99$ c) $P=5.2$ d) $R=1.6216$

Section C

DIAGRAMMATIC REPRESENTATION AND STATISTICS

Please note that questions are not included with the data in this section, but at the beginning of the section there is a list of specific objectives and a list of the applicable data.

When you have completed this section you should be able to:

1. Use parallel scales to convert one set of data to another system of related units.
2. Plot and correctly label a graph from given values.
3. Read values from a graph.
4. Determine the gradient of a straight line graph.
5. Tabulate and group statistical data.
6. Produce a tally, frequency diagram, 100% bar chart, pie diagram, pictogram, bar chart and histogram.
7. Interpret data represented by **6**.

1. Produce and use parallel scales for converting to another system of related units.

   USE Q 2.1(a) to (d)

   i.e. construct parallel scales for 2.1(a) and hence convert $10\frac{1}{2}$ hours into minutes.

2. Plot a graph and use to find corresponding pairs of values.

   USE Q 2.2 and 3.1

   i.e. construct a graph of $y = 3x - 2$ between $x = 2$ and $x = 8$ and use to determine the value of x when $y = 15\frac{1}{2}$.

3. Plot physical quantities which are directly proportional and find corresponding pairs of values.

   USE Q 2.3 to 2.9

   i.e. plot a graph of the values in Q 2.3 and use the graph to determine the mass required to cause an extension of 130 mm.

4. Plot physical quantities of non-linear graphs and use to find corresponding pairs of values.

   USE Q 3.2 to 3.8

   i.e. plot a graph of the values in Q 3.2 and use to determine the average age of boys weighing 60 kg.

5. Plot, label, and find the gradient of a graph which follows a linear law.

   USE Q 2.2 to 2.9

   i.e. plot a graph of the values in Q 2.7 and find the relationship between V and F.

6. Group data and produce a tally.

   USE Q 4.7, 4.8, 4.11, and 4.14–4.17

7. Draw a:

   a) 100% bar chart USE Q 4.6, 4.10, 4.12
   b) Pie diagram USE Q 4.6, 4.10, 4.12
   c) Pictogram USE Q 4.1, 4.2, 4.4, 4.6
   d) Histogram USE Q 4.1, 4.3, 4.5, 4.9, etc.
   e) Frequency table USE Q 4.7, 4.8, 4.11, 4.14–4.17
   f) Bar chart USE Q 4.1, 4.2, 4.4, 4.5, etc.

1. Construct parallel scales (or a graph) to permit conversion of the following:
   a) Hours to minutes (range 0 to 12 hours).
   b) Fahrenheit to Celsius (range -100 to $+250\,^\circ$C).
   c) Miles to kilometres (range 0 to 100 miles) (1 mile = 1.609 km).
   d) Inches to millimetres (range 0 to 20 inches) (1 inch = 25.4 mm).

*2. a) $y = 4x$ b) $y = 3x - 2$ c) $y = 4 - 2x$ d) $3y = 2x - 1$

3. The table below gives a set of readings obtained when a mass of W(g) is hung from a spring, causing it to extend to a length L (mm):

W	10	40	80	100	120	150
L	116	123	144	152	160	173

4. The electrical resistance of various lengths of wire are given in the following table:

Length (mm)	110	120	150	160	190	210
Resistance (ohms)	3.3	3.6	4.5	4.8	5.7	6.3

5. In an experiment a quantity of air was heated and its volume changed as indicated below:

Temp. °C	22	68	100	125	155	180	204	250
Volume ($m^3 \times 10^{-6}$)	59.0	68.2	74.6	79.9	80.3	91.0	95.4	104.6

Note: There is an error in one of these readings.

6. The following is a comparison table between Celsius and Fahrenheit temperature scales:

C	10	25	50	60	75	100
F	50	77	122	140	167	212

7. The relationship between force and velocity for moving a certain object is indicated below:

F	1.9	3.3	6	8	9.3
V	1.6	4	8.6	12	14.2

8. When current is taken from a primary cell the internal resistance of the cell causes a drop in terminal voltage. The following is the result of a test:

Output current	I	0.1	0.3	0.5	0.7	0.9	1.0
Terminal voltage	E	1.44	1.32	1.2	1.08	0.96	0.9

9. The table shows the length of a metal rod at different temperatures. It is assumed that the relationship is $L = a + bt$:

t (°C)	0	10	20	30	40	50	60
L (mm)	100	100.02	100.04	100.06	100.08	100.10	100.12

The laws applicable to Qs 3–9 are given on P 64.

1. Plot the graph of:

a) $y = x^2 + x$ from $x = -4$ to $+4$

b) $y = 2x^2 - 3x$ from $x = -4$ to $+4$

c) $y = 3x^2 - 7x + 6$ from $x = -4$ to $+4$

d) $y = \frac{1}{x}$ from $x = 0$ to 5

e) $y = \sin\theta$ from $\theta = 0$ to $90°$

f) $y = \log x$ from $x = 1$ to 10

2. The average weight of boys of different ages are given in the following table:

Age in years	11	12	13	14	15
Weight – kg	40	42.5	46	50.5	57

3. The following table gives the distance travelled from rest after a given time:

Time in seconds	0	3	6	9	12	15
Distance in metres	0	2	8	18	32	50

4. The following table gives the results of a test on a Pelton wheel:

Rev/min	200	235	280	335	470	650	740	800
% efficiency	26	40	55.5	68	85	86	76	64

5. The breaking load of different steel wires are given in the table according to wire circumference:

Load	4	9	16	25	36	49	64
Circumference	1	1.5	2	2.5	3	3.5	4

6. Weight per unit length of round aluminium bar varies with the diameter according to the following table:

Diameter (mm)	10	20	30	35	40
Weight (N/m)	2	9	19	26	35

7. The Highway Code records that the shortest distance required to stop a vehicle after seeing an emergency is as follows:

Miles/hour	20	30	40	50	60	70
Distance (m)	13	25	40	56	80	105

8. The mass of a baby was recorded every two weeks for the first 16 weeks of life and recorded as follows:

Age in weeks	0	2	4	6	8	10	12	14	16
Mass in kg	3.2	3.3	3.6	3.9	4.1	4.1	4.4	4.9	5.3

1. During a 'flu epidemic the number of people absent on ten consecutive days were as follows:

Day	1	2	3	4	5	6	7	8	9	10
No.	30	35	60	78	87	83	69	43	30	29

2. The numbers of people employed by 6 firms in a certain area are:

Firm	1	2	3	4	5	6
Employees	750	160	300	1700	550	900

3. Sales of a retail store selling heating appliances:

Month	May	June	July	Aug	Sept	Oct	Nov	Dec	Jan	Feb	Mar	Apr
Sales	82	60	52	64	104	130	160	186	152	144	104	86

4. The number of vehicles per kilometre of road in certain countries in 1968 was estimated to be:

India	1	Japan	13	Italy	31
Australia	5	U.S.A.	16	Netherlands	32
Canada	10	France	16	West Germany	32
New Zealand	11	Switzerland	22	Great Britain	38

5. The following gives the number of passes to the nearest 100 pupils, who obtained 1, 2, 3, . . . 7 passes in Higher Grade subjects in the 1971 C.S.E. examinations:

Subjects passed	1	2	3	4	5	6	7
Number of candidates	4900	4800	3900	3400	2900	1100	100

6. The percentage of leisure time used for various activities of a particular group of students is given below:

Television	24%	Reading	14%
Hobbies and Crafts	11%	Walking	8%
Gardening	9%	Social	12%
Sport	12%	Clubs	10%

7. The table below gives the diameters of a batch of balls to the nearest mm.

78	72	74	74	79	71	75	74
72	73	73	72	74	75	74	71
66	75	71	73	74	72	79	74
80	69	74	70	70	80	75	76
72	73	70	77	68	72	75	67

8. The weekly earnings of 50 men in a factory are given below:

87.50	86.75	91.00	88.80	87.85
86.35	86.75	88.00	89.25	87.75
87.65	86.90	87.00	88.25	88.25
90.00	89.20	89.00	88.80	89.65
89.50	86.80	87.50	88.00	87.50
88.00	90.85	88.50	89.25	88.75
86.35	88.00	88.65	87.75	88.25
87.85	89.00	87.50	89.50	88.00
89.20	90.00	88.20	87.65	88.00
90.00	87.25	87.50	88.50	90.25

9. The following gives the frequency of telephone calls of different durations (to the nearest minute) through a particular exchange:

Duration (mins)	0	1	2	3	4	5	6	7	8
Frequency	3	20	31	35	28	15	5	8	2

10. The income of a particular family is £90 per week of which £30 is deducted at source for tax, etc., £15 is needed to pay off the mortgage and £20 is used to buy food. Of the remainder, £10 is put aside for annual bills and the rest is split equally between clothing, entertainment and savings for a holiday.

11. The marks obtained by 30 candidates in an examination in which the lowest mark possible was 1 and the highest was 50 were as follows:

38	6	25	18	23	49
30	40	28	43	35	48
21	11	27	13	28	27
26	33	32	45	34	23
16	18	24	30	27	19

12. A company surveyed the mode of travel of its employees before and after introducing a company bus service. The results are as follows:

	Train	Car	Motor cycle	Cycle	Walking	Company bus
Before %	10	35	20	25	10	0
After %	5	25	15	15	10	30

13. The height registered by a barometer was recorded on 100 occasions as follows:

Height mm	740	741	742	743	744	745	746	747	748	749	750
Frequency	1	5	9	23	20	17	12	6	4	2	1

14. The following table records the lateness in minutes of a group of 100 students for a particular lecture:

0	7	0	7	0	0	0	9	0	0
1	0	0	0	0	9	0	0	0	1
6	11	14	0	3	0	2	0	3	0
0	0	2	20	0	0	0	0	0	0
0	5	0	0	0	0	28	5	0	0
4	0	0	8	0	3	0	0	13	10
15	0	16	0	0	0	4	0	0	0
0	6	0	0	5	0	0	2	10	3
0	23	12	4	0	18	0	0	0	0
1	0	0	0	0	0	8	13	0	0

15. The marks attained by a group of 80 students in a mid-session test were recorded as follows:

31	54	43	52	64	46	55	66	36	46
15	32	21	65	23	53	75	27	57	65
69	99	53	33	54	14	52	44	74	98
88	46	66	07	42	38	25	100	89	53
27	57	83	47	84	61	43	31	61	74
51	68	16	79	38	75	82	52	26	39
72	38	59	25	52	43	66	79	48	67
42	41	37	91	44	79	31	67	58	45

16. In order to test the reliability of a particular method of determining temperature, the same temperature was determined by 30 different people, with the following results (temperature in °C):

20.1	19.7	20.7	20.3	20.3	19.2	20.1	21.2	19.8	20.5
20.9	20.6	21.6	18.7	21.4	21.7	22.1	19.3	21.8	21.3
19.6	20.2	19.9	21.1	19.8	20.8	20.4	20.9	20.9	19.4

17. The weight of 40 glass milk bottles were determined accurately in order to calibrate a bottle-filling machine. The results (in kg) are recorded below:

0.358	0.359	0.360	0.360	0.358	0.360	0.359	0.360
0.360	0.361	0.358	0.361	0.360	0.357	0.360	0.358
0.356	0.357	0.359	0.357	0.359	0.361	0.361	0.359
0.361	0.358	0.360	0.358	0.362	0.359	0.362	0.359
0.362	0.361	0.359	0.359	0.360	0.358	0.358	0.363

1. Construct a parallel scale to convert pounds weight to kilograms over a range of 10 lb. Using your scale convert 6 lb 12 oz to kg. (1 lb = 16 oz). Use 1 lb = 0.454 kg.

2. Plot the graph of $4y = 2x + 5$ between $x = -5$ to $+5$ and from your graph determine x when $y = 2$.

3. In an experiment to determine the friction between two metallic surfaces the following values of load and effort were obtained:

Load (N)	30	50	70	100	120
Effort (N)	6.2	15	23	36	44

Plot these values (load horizontally), read from the graph the load when the effort is 20 N and find the gradient of the graph.

4. The velocity of a rocket as it leaves the ground is recorded below:

Time (secs)	0	1	2	3	4	5	6	7
Velocity (m/sec)	0	5	18	38	62	76	81	83

Plot this information and estimate from your graph:

a) the time to reach a velocity of 45 m/sec,

b) the velocity after 8 sec.

5. The following shows how the county spend each pound of the rates collected:

Housing	7 p	Police and Fire	12 p
Education	56 p	Social Services	7 p
Refuse, etc.	5 p	Other expenses	4 p
Highways	9 p		

Illustrate this information by:

a) a pie diagram b) a 100% bar chart.

6. 30 – 5 Ω resistances were measured accurately at 18 °C with the following results. Produce a tally and hence a frequency table and illustrate by a histogram.

5.02	5.02	5.00	5.02	5.01	5.01
5.00	5.01	5.02	5.00	5.00	5.00
5.01	4.99	5.01	5.00	5.03	5.00
5.00	5.00	4.98	5.01	5.00	5.05
4.99	4.98	5.01	4.99	5.00	5.03

(Answers on P 64)

Section D

GEOMETRY, TRIGONOMETRY AND MENSURATION

When you have completed this section you should be able to:

1. Use the relationships associated with parallel and transverse lines.
2. Recognise types of triangles and their properties.
3. Construct triangles.
4. Apply the theorem of Pythagoras.
5. Use the geometric properties of a circle.
6. Convert degrees to radians and vice versa.
7. Calculate the areas of plane geometric figures.
8. Calculate the surface area and volume of geometric solids.
9. Convert to and from the three basic trigonometrical ratios using tables.
10. Solve right angled triangles using the three basic trigonometrical ratios.

1. Determine the unknown angles α, β and γ in the diagram Fig. Q1.

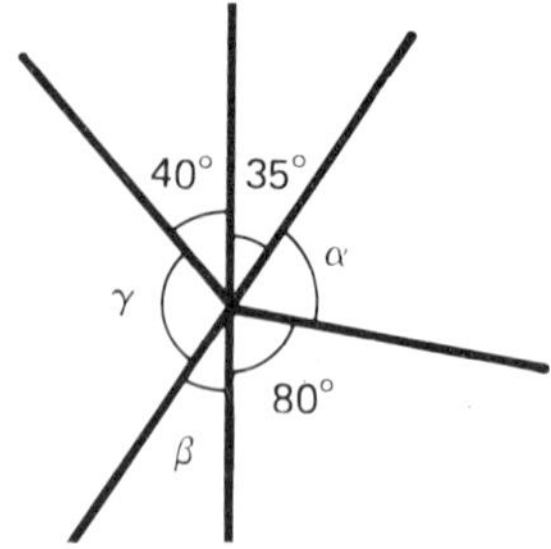

Fig. Q1

2. a) Find the complementary angles to 49°, 37° 21′, 79° 6′.
   b) Find the supplementary angles to 78°, 80° 12′, 147° 49′.

3. AOD is a straight line. If B and C are points on the same side of the line such that ∠BOC = 90° and ∠BOA = 30° determine the value of ∠COD.

4. AOD is a straight line and line CO is drawn such that ∠COA = 55°. If the reflex angle COB is bisected by line DO, find ∠DOB.

5. EF is a line cutting two parallel lines AB and CD at points G and H respectively. If ∠EGB is 65° find all remaining angles.

6. A gear wheel has 40 teeth. Find the angle of rotation when:
   a) 4 teeth b) 25 teeth have passed a given mark.

7. ABCD is a parallelogram with ∠ADC = 58°. Find the remaining interior angles.

8. Find the unknown angles in the diagrams Fig. Q8.

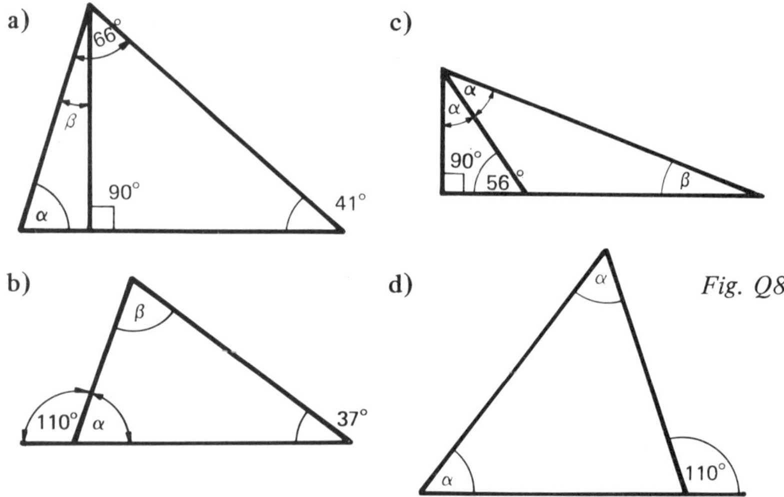

Fig. Q8

9. The end view of a roof is triangular in which the angle at the apex is 120°. If the sides are equal, what angle do they make with the horizontal?

10. ABC is a triangle with ∠A = 58°. A line DC is drawn perpendicular to BC such that ∠ACD = 34°. Find the value of ∠ABC.

11. A wall crane consists of tie AB and strut CB hinged to a vertical wall at A and C. If the tie is at 45° to the wall and the strut at 30° to the wall find the angle ABC.

12. If one triangle has sides of 6 cm (base) 4 cm and 5 cm and a similar triangle has a base of 15 cm find the lengths of its other sides.

13. A straight line parallel to base BC of triangle ABC meets AB at X and AC at Y.
If AB = 12, AC = 8, BC = 10 and AX = 9 calculate lengths of sides AY and XY.

14. A lamp 9 m high is placed 6 m from a wall 5 m high.
Find the length of the shadow of the wall.

15. In triangle ABC $\angle B = 90°$:

a) If $a = 10$ and $c = 14$ find b
b) If $a = 3$ and $b = 6$ find c

16. A ladder 12.5 m long touches a vertical wall 11 m from the ground. How far is the foot of the ladder from the wall?

17. Two posts 30 m apart are on level ground. If the posts are 6 m and 10 m high respectively what is the distance between their tops?

18. 3 holes, A, B and C are drilled in a plate on the circumference of a circle centre O.
If AB = AC and $\angle$BOC is 76° find the angles of the triangle ABC.

19. 3 transistors A, B and C are placed on a board on the circumference of a circle such that BC is a diameter. If AB = 0.06 m and BC = 0.10 m calculate the length of wire to join AB and C.

20. A flat surface of width 180 mm is to be machined on a shaft 260 mm dia.
Find the depth of material removed.

21. Find the angles A, B and C in Fig. Q21.

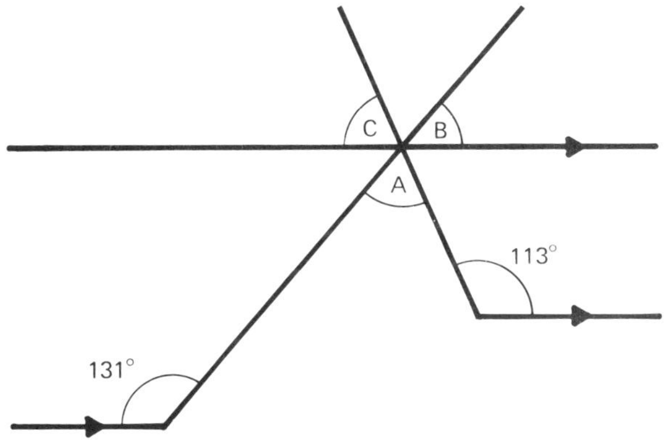

Fig. Q21

22. Two sides of a triangle are 23 mm long, and the angle between them is 62° 30′.
Determine the remaining angles of the triangle.

23. If the longest side of a right angled isosceles triangle is 120 mm calculate the lengths of the other sides.

24. Calculate the smallest diameter hole a rectangular bar 12 mm by 24 mm will pass through.

1. A clock started at noon. Through what angles will the minute hand have turned by:
   a) 2.45 p.m.,
   b) ten minutes past 4,
   c) what will be the time after 192°?
2. x° is the smallest angle in a triangle. If the others are $2.5x^\circ$ and $4.5x^\circ$ find each angle.
3. The diameter AB of a semicircle is 2.5 m. A point P on the circle is 1.8 m from A. How far is P from B?
4. A chord of a circle is 2.6 m long and 1.1 m from the centre. Find the radius of the circle.
5. A regular hexagon has a side length of 2.6 mm. What is:
   a) the distance across opposite corners of the hexagon,
   b) the distance across opposite flats of the hexagon.
6. One external angle of a right angled triangle is $102^\circ\ 37'\ 25''$. What are the two internal angles?
7. A chord AB is drawn so that the angle in the minor segment of Fig. Q7 is 108°.
   If angle ABC = 90° what are angles CAB and ACB?

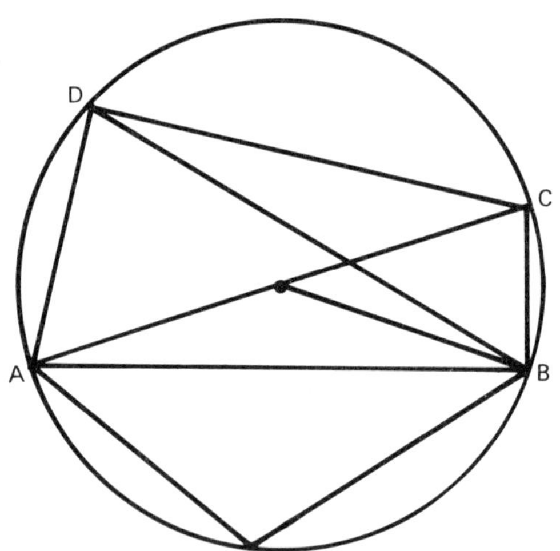

Fig. Q7

8. In Fig. Q7 find also ADC and angle ADB.
9. If AD = 375 mm (Fig. Q7) and angle DAC = 60° what is the diameter of the circle?

10. Construct a triangle ABC given:

a) $a = 40$ mm, $b = 50$ mm, $c = 60$ mm
b) $a = 60$ mm, $b = 70$ mm, $\angle C = 50^\circ$
c) $a = 80$ mm, $\angle B = 30^\circ$, $\angle C = 40^\circ$
d) $a = 70$ mm, $\angle A = 90^\circ$, $c = 40$ mm

11. a) Add (i) $24^\circ\ 37'$ to $46^\circ\ 52'$
(ii) $35^\circ\ 21'\ 35''$ to $66^\circ\ 41'\ 40''$

b) Subtract (i) $57^\circ\ 35'$ from $65^\circ\ 24'$
(ii) $24^\circ\ 21'\ 46''$ from $43^\circ\ 18'\ 31''$

c) Convert (i) $63^\circ\ 21'\ 40''$ to decimals of a degree
(ii) 49.468° to degrees, minutes and seconds

12. Change the following degrees to radians in an exact form (using π):

a) 90° b) 45° c) 75° d) $52^\circ\ 30'$ e) $72^\circ\ 30'$

13. Change the following radians to degrees:

a) $\frac{\pi}{3}$ b) $\frac{2\pi}{5}$ c) $\frac{3\pi}{6}$ d) $\frac{2\pi}{9}$ e) $\frac{\pi}{18}$

14. Express the following radians in degrees and minutes using four figure tables:

a) 0.2 b) 0.62 c) 1.45 d) 0.2049 e) 0.5585

15. Change the following degrees and minutes to radians using four figure tables:

a) 35° b) $47^\circ\ 30'$ c) $67^\circ\ 20'$ d) $88^\circ\ 45'$ e) $17^\circ\ 12'\ 20''$

16. A tangent is drawn from point B on a circle to distant point A 90 mm away.
If the diameter of the circle is 50 mm find the distance from A to the centre of the circle.

17. The angle subtended at the centre of a 89 mm diameter circle is 4.5 radians.
Find the lengths of the major and minor arcs.

18. How many revolutions will a 700 mm diameter bicycle wheel make in travelling 1.2 km?

1. In triangle ABC determine:
   a) sine ratio of angle B,
   b) cosine of angle A,
   c) sine of angle A,
   d) tangent of angle B.

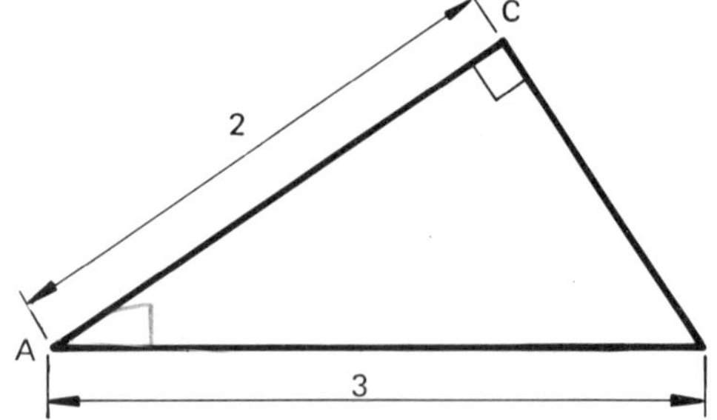

Fig. Q1

2. Use tables to find the *tangent* of the following angles:

a) 61°	c) $34^\circ 26'$	e) $51^\circ 23'$
b) $63^\circ 42'$	d) $87^\circ 30'$	f) $55^\circ 8' 20''$

3. Use tables to find the *angles* whose *tangents* are (*arctan* of):

a) 1.3270	c) 1.7427	e) 0.7329
b) 0.1817	d) 0.2529	f) 0.7309

4. Find the *sine* of:

a) 33°	c) $63^\circ 35'$	e) $2^\circ 30'$
b) $14^\circ 24'$	d) $87^\circ 30'$	f) $32^\circ 22' 30''$

5. Find the angle whose *sine* is (*arcsin* of):

a) 0.3665	c) 0.1985	e) 0.0265
b) 0.6497	d) 0.3126	f) 0.7409

6. Write down the *cosines* of the following angles:

a) 85°	c) $75^\circ 13'$	e) $70^\circ 11'$
b) $24^\circ 30'$	d) $15^\circ 59'$	f) $80^\circ 25' 20''$

7. Use tables to find the *angles* whose *cosines* are (*arccos* of):

a) 0.0872	c) 0.9030	e) 1.7601
b) 0.7749	d) 0.0055	f) 0.8880

8. *Without* using tables write down the following:

a) the *sine* of 30°	c) *cosine* of 90°
b) the *tangent* of 45°	d) *tangent* of 60°

9. Without using trigonometrical tables find:

a) $\cos 30^\circ$	c) the angle whose *sine* $= \frac{1}{2} \tan 45^\circ$
b) $\sin 45^\circ$	d) the angle whose *cosine* $= \sin 90^\circ - \tan 45^\circ$

1. ABC is a right-angled triangle with angle B = 90°.
   a) If $a = 4.8$ and $b = 7.6$ find angle A.
   b) If $a = 6.7$ and $b = 9.4$ find angle C.
   c) If $a = 2.9$ and $c = 1.9$ find angle A.
2. In a right-angled triangle the side adjacent to one of the acute angles and the right angle is 16.9 mm and the side opposite this acute angle is 16.2 mm.
   What are the angles of the triangle?
3. ABC is a right-angled triangle.
   a) If $\angle B = 90°$, $\angle A = 40°$ and $a = 24.3$ find sides b and c.
   b) If $\angle A = 90°$, $\angle B = 74° 29'$ and $c = 1.48$ find the other sides.
   c) If $\angle A = 90°$, $\angle C = 62° 16'$ and $a = 3.59$ find the other sides.
4. In a right-angled triangle the hypotenuse is 6.33 m long and one of the angles is 33° 30′.
   What is the length of the side opposite the angle?
5. ABC is a right-angled triangle with angle A = 90° and AD is a line perpendicular to BC. If $\angle ACB = 35°$ and AB = 12 mm, calculate the lengths of AC and DC.
6. In triangle ABC angle B = 90° and a perpendicular is drawn from B to AC meeting it at D. If $\angle C = 64°$ and AB = 72 mm find the length of AD.
7. An isosceles triangle has equal sides of length 45 m and a base length of 32 m.
   What are the angles of the triangle?
8. A rhombus has sides of length 400 mm. If the angle between two of the sides is 62°, find the lengths of the diagonals. (*Note*: diagonals of a rhombus intersect at 90°.)
9. In a parallelogram ABCD the diagonal AC is perpendicular to side BC. If BC = 70 mm and $\angle BAC = 32°$ calculate the length of side AB.
10. ABC is an isosceles triangle with AB = AC = 142 mm and angle ABC is 58°. D is a point on AB such that the line CD is perpendicular to AB.
    Find the length of CD.
11. Find the total length of strip required to make the bracket shown in Fig. Q11.

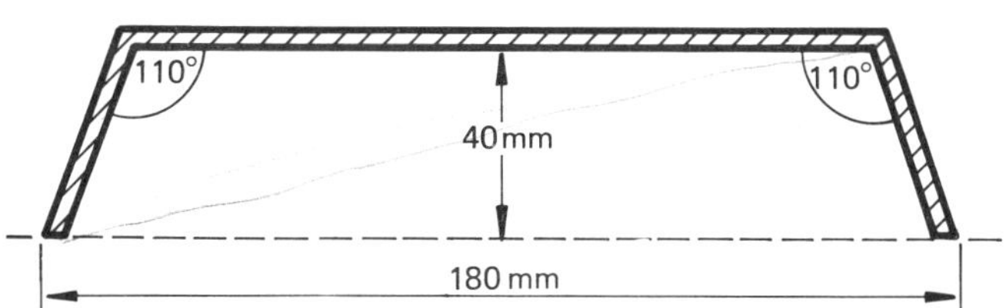

Fig. Q11

12. A boat leaves harbour at 12 km/hr on a bearing N 32° W. After $\frac{3}{4}$ hr its bearing changes to N 48° E.
    How far is it from the harbour after a total of $1\frac{1}{2}$ hours?

1. A gate $2\frac{1}{2}$ m long by $1\frac{1}{2}$ m high is strengthened by a diagonal bar. What is the length of the bar, and what angle does it make to the bottom rail?
2. A road has a steady incline of $10^\circ\ 15'$ over 200 m horizontal distance. How high has it climbed?
3. Each step on a set of stairs has a rise of 190 mm and a tread of 250 mm.
   What angle does the hand rail make with the horizontal?
4. A point 100 m from the centre of a factory chimney supports a theodolite of $1\frac{1}{2}$ m height. From there the theodolite gives an angle of $14^\circ\ 48'$ to the top of the chimney.
   What is the height of the chimney?
5. A pair of dividers 120 mm long is open to measure 40 mm.
   What is the angle between the legs?
6. A 10 m ladder just reaches the gutter 8 m above the ground.
   What angle does the ladder make with the ground and how far is its foot from the house?
7. A 700 mm bicycle wheel has a chalk mark on the tyre. If the mark is touching the road before moving the wheel 40° how high will the mark rise above the road?
8. A symmetrical roof is shown in the diagram, find its greatest height.

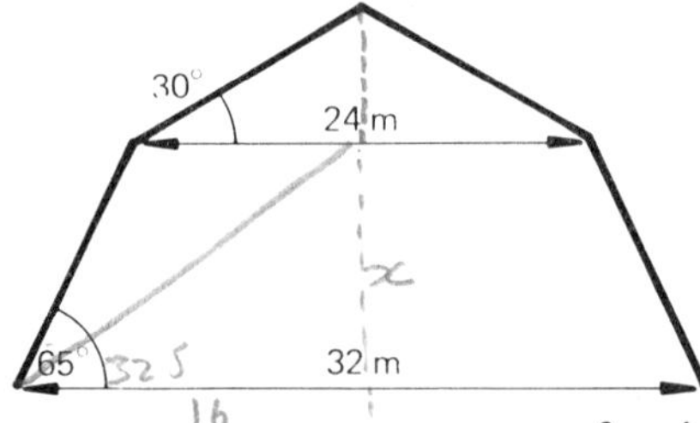

Fig. Q8

9. The altitude of the sun was known to be $34^\circ\ 13'$. The length of the shadow of a chimney was 55 m.
   What is the height of the chimney?
10. A pilot flying at 1000 m observes a bridge at an angle of depression of $15^\circ\ 35'$. Ten seconds later the angle is $22^\circ\ 10'$.
    If the aeroplane is flying towards the bridge what is its speed? (in m/s).
11. Without using tables write down the following values related to Fig. Q11:
    a) DC
    b) BD
    c) AD
    d) AB
    e) $\cos 60^\circ$
    f) $\tan 45^\circ$

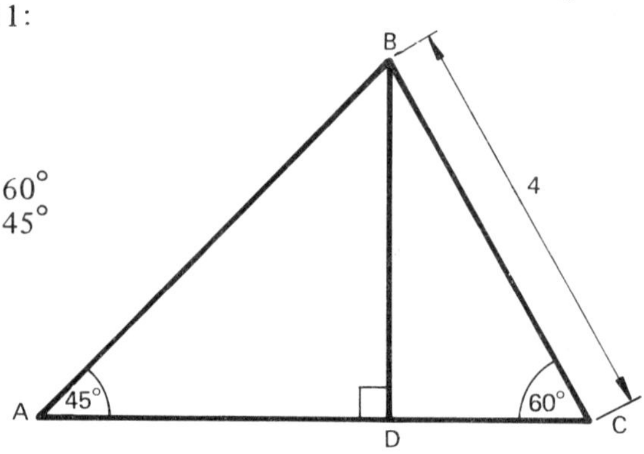

Fig. Q11

1. Calculate the area of:
   a) a *square* measuring 25 mm × 25 mm,
   b) a *rectangle* measuring 35 mm × 75 mm,
   c) a *parallelogram* measuring 30 mm × 50 mm at 60°,
   d) a *circle* having a diameter of 56 mm if $\pi = \frac{22}{7}$,
   e) a *semi-circle* having a diameter of 84 mm if $\pi = \frac{22}{7}$.

2. A rectangular sheet measures 1.5 m × 2.5 m, what is its area?

3. A metal plate 90 mm wide has an area of 14 040 mm^2, find its length.

4. Calculate the area of a parallelogram whose base is 47.3 mm and whose vertical height is 23.5 mm.

5. A 45° parallelogram has an area of 3.75×10^5 and a base length of 750 mm.
   Find the length of the shorter side.

6. A rhombus having a 60° angle has a side length of 85 mm.
   Calculate its area.

7. A trapezium has parallel sides of 85 mm and 120 mm, 47 mm apart.
   Find the area.

8. A circle has a diameter of 756 mm, calculate its area using $\pi = 3.142$.

9. Calculate the area of a ring 86 mm outside diameter and 74.5 mm inside diameter.

10. A sector of a circle has an angle of 90° and a radius of 56 mm, calculate its area and the area of the segment if the triangle is removed.

11. A circle has an area of 707 mm^2, calculate its diameter ($\pi = 3.142$).

12. An annulus has an outside diameter of 49 mm and an area of 1709 mm^2. Calculate its internal diameter ($\pi = 3.142$).

13. Calculate the areas of the following sectors:
    a) A radius of 56 mm subtended at the centre by an angle of 48°.
    b) A radius of 0.14 m subtended at the centre by an angle of 107°.
    c) A radius of 165 mm subtended at the centre by an angle of 2.1 radians.

14. Calculate the perimeter of the three sectors referred to in question **13**.

15. A rectangle is twice as long as its width and has an area of 700 m^2. Calculate the length of one diagonal.

1. Calculate the volume of:
   a) a *cylinder* having a diameter of 54 mm and length 86 mm,
   b) a *prism* having a rectangular base 36 × 18 and 72 mm height,
   c) a *cone* having a 50 mm diameter base and 75 mm height,
   d) a *pyramid* having a 95 mm square base and 120 mm height,
   e) a *sphere* having a diameter of 100 mm.

2. Calculate the surface area of:
   a) a *cylinder* 95 mm diameter by 150 mm long,
   b) a *rectangular prism* 84 × 64 × 38 mm.

3. A rectangular fish tank 350 × 350 × 900 mm long contains water to a depth of 300 mm.
   Calculate the volume of water.

4. A concrete base is to be 2 m × 3 m × 250 mm thick.
   Calculate the volume of concrete required.

5. Calculate the surface area of an open topped cylinder 0.8 m diameter × 1.2 m height.

6. Calculate the volume of oil contained in a drum 0.85 in diameter × 1.5 m.

7. A cylinder 69 mm high has a volume of 3.48×10^6 mm^3, find its diameter.

8. A conical hopper has a diameter of 0.92 m and a perpendicular height of 1.25 m, calculate its volume.

9. An open trough 2 m long with a cross section 300 mm deep is 500 mm wide at the top and 400 mm wide at the bottom.
   Calculate the volume and surface area of sheet metal involved.
   The trough has no ends.

10. A hemisphere has a volume of 1.2×10^4 mm^3, find its diameter.

11. An ingot 80 × 10 × 300 mm long is cast into a cylinder 120 mm diameter.
    Calculate its length and total surface area.

12. A rivet has a hemispherical head 6 mm radius and a stem 6 mm diameter by 15 mm long.
    Find the volume occupied by 100 such rivets.

13. An extrusion has a cross sectional shape of a sector of a circle 25 mm radius with a subtended angle of 50°.
    If the extrusion is 3.2 m long, what is the volume?

14. If the extrusion referred to in question **13** is cut into 50 mm lengths, calculate the surface area of each piece.

1. A template is in the form of a parallelogram. The angles at the corners being 55° and 125° and the sides measure 17.5 mm and 32.5 mm.
   Find the area.

2. An extruded bar has a cross-section in the form of an equilateral triangle of 25 mm side length.
   Find the cross-sectional area.

3. A specimen of 17.5 × 7.5 mm rectangular cross-section was subject to a tensile stress until it fractured.
   If the area of fracture was 95 mm^2 calculate the percentage reduction in area.

4. A rectangular bench 810 mm × 380 mm has rounded corners of 47.5 mm radius.
   Calculate its surface area.

5. The surface area of a cylindrical tank (whose diameter and height are equal) is known to be 5.564 m^2.
   Assuming no lid, find the diameter.

6. The development of the surface for a conical sheet metal hood is a sector of a circle. Its area is 16 900 mm and it subtends an angle of 120°.
   What is the slant height, vertical height and diameter of the cone?

7. A bar of iron 160 × 40 × 20 mm was heated and forged until its cross-sectional area was 35 × 7.5 mm.
   How long has it become?

8. A rod of copper 260 mm long × 12.5 mm diameter is extruded into wire 0.5 mm diameter.
   What length of wire will be obtained?

9. A spherical brass weight is to be cast to have a mass of 3.2 kg.
   Calculate its diameter if the density of brass is 8.9 g/cm^3.

10. A lead cone was cast from a rectangular ingot 160 × 25 × 17.5 mm.
    If the diameter of the base of the cone was 102.6 mm what was the vertical height?

11. A round bar 100 mm long has a volume of 8×10^6 mm^3.
    Determine its diameter and hence the surface area.

12. A metal clip is in the shape of a letter 'C', outer diameter 10 mm, inside diameter 2 mm. The portion cut away being a sector subtending an angle of 60° to the centre.
    Find the area.

1. a) Construct triangle ABC if $a = 80$ mm, $c = 50$ mm, $\angle A = 90°$.
   b) Calculate the length of side b of the above triangle.
   c) Calculate the area of the above triangle.
2. a) In triangle XYZ angle $X = 27° 43' 27''$ and angle $Y = 87° 51' 40''$ find angle Z.
   b) Convert angle X above into decimals of a degree.
   c) Convert angle Y above into radians.
3. If the angle subtended at the centre of a circle 120 mm diameter is 1.8 radians:
   a) find the length of the minor arc,
   b) find the circumference of the circle,
   c) find one of the identical angles in the isosceles triangle thus formed (in degrees).
4. a) Find the sine, cosine and tangent of (i) $23° 27'$, (ii) $57° 14' 30''$.
   b) In triangle ABC, $\angle A = 90°$, $\angle B = 37° 27'$, $c = 54$ mm, find the unknown sides and angle.
   c) Calculate the area of the above triangle.
5. An oil rig is viewed from the top of a cliff 256 m high. The angle of depression of the top of the rig is $20° 21'$ whilst to the rig at sea level it is $31°$.
   How high does the rig stand above sea level?
6. A steel spacer ring 0.2 m thick has an outside diameter of 1.1 m and an inside diameter of 0.8 m:
   a) Find the volume of steel involved.
   b) Find the total surface area.
7. A solid plumb bob is composed of a hemisphere and cone of equal diameter of 20 mm. The total length is 40 mm:
   a) Find the total volume of the solid.
   b) Find the length of ingot 10 mm × 20 mm cross-section from which the solid can be cast.
8. A triangle is drawn within a circle such that each corner touches the circle and one side passes through the centre of the circle.
   If the shortest side of the triangle is half the length of the longest side what are the angles of the triangle?
9. Calculate the area of the largest and smallest segments formed by the triangle in question 8 if the diameter is 100 mm.
10. In triangle ABC angle B is $90°$, $a = 37$ mm and $b = 42$ mm. Calculate angles A and C in radians.

GEOMETRY 1

1. 65°, 35°, 105°
2. a) 41°, 52° 39′, 10° 54′ **b)** 102°, 99° 48′, 32° 11′
3. 60° **4.** 125° **5.** 65°, 115°
6. a) 36° **b)** 225° **7.** 58°, 122°, 122°
8. a) 17°, 73° **b)** 73°, 70° **c)** 34°, 22° **d)** 55°, 55°
9. 30° **10.** 66° **11.** 15°
12. 12.5 cm, 10 cm **13.** 6, 7.5 **14.** 7.5 m
15. a) 17.2 **b)** 5.196 **16.** 5.94 m **17.** 30.265 m
18. 71°, 71°, 38° **19.** 0.24 m **20.** 36.19 mm
21. 64°, 49°, 67° **22.** 58° 45′ **23.** 84.85 mm
24. 26.83 mm

GEOMETRY 2

1. a) 990° **b)** 1500° **c)** 12.32 hr
2. 22° 30′, 56° 15′, 101° 15′ **3.** 1.735 m
4. 1.703 m **5. a)** 5.2 m **b)** 4.503 m
6. 77° 22′ 35″, 12° 37′ 25″ **7.** 18°, 72°
8. 90°, 72° **9.** 750 mm **10.** No answers
11. a) 71° 29′, 102° 3′ 15″ **b)** 7° 49′, 18° 56′ 45″ **c)** 63.361″, 49° 28′ 4.8″
12. a) $\frac{\pi}{2}$ **b)** $\frac{\pi}{4}$ **c)** $\frac{5\pi}{12}$ **d)** $\frac{7\pi}{24}$ **e)** $\frac{29\pi}{72}$
13. a) 60° **b)** 72° **c)** 90° **d)** 40° **e)** 10°
14. a) 11° 28′ **b)** 35° 31′ **c)** 83° 5′ **d)** 11° 44′ **e)** 32°
15. a) 0.6109 **b)** 0.8290 **c)** 1.1752 **d)** 1.5490 **e)** 0.3003
16. 93.41 mm **17.** 200.25 mm, 79.39 mm
18. 545.6 revolutions

TRIGONOMETRY 3

1. a) $\frac{2}{3}$ **b)** $\frac{2}{3}$ **c)** $\frac{\sqrt{5}}{3}$ **d)** $\frac{2}{\sqrt{5}}$
2. a) 1.804 **b)** 2.0233 **c)** 0.6856 **d)** 22.90 **e)** 1.2519 **f)** 1.4356
3. a) 53° **b)** 10° 18′ **c)** 60° 9′ **d)** 14° 11′ 40″ **e)** 36° 14′ 12″
f) 36° 9′ 45″
4. a) 0.5446 **b)** 0.2487 **c)** 0.8956 **d)** 0.9990 **e)** 0.0436 **f)** 0.5355
5. a) 21° 30′ **b)** 40° 31′ 30″ **c)** 11° 26′ 40″ **d)** 18° 13′ **e)** 1° 31′
f) 47° 48′ 30″
6. a) 0.0872 **b)** 0.9100 **c)** 0.2551 **d)** 0.9613 **e)** 0.3390 **f)** 0.1664
7. a) 85° **b)** 39° 12′ **c)** 25° 26′ **d)** 89° 41′ **e)** Impossible
f) 27° 22′ 30″
8. a) 0.5 **b)** 1 **c)** 0 **d)** $\sqrt{3}$
9. a) $\frac{\sqrt{3}}{2}$ **b)** $\frac{1}{\sqrt{2}}$ **c)** 30° **d)** 90°

TRIGONOMETRY 4

1. a) 39° 10′ **b)** 44° 32′ **c)** 56° 46′ **2.** 43° 47′ 20″, 46° 12′ 40″
3. a) 37.8, 28.96 **b)** 5.532, 5.331 **c)** 1.671, 3.178
4. 3.494 m **5.** 17.14 mm, 14.04 mm
6. 64.71 mm **7.** 69° 10′, 41° 40′ **8.** 685.8 mm, 412 mm
9. 132.1 mm **10.** 127.6 mm **11.** 236 mm
12. 13.79 km

TRIGONOMETRY 5

1. 2.915 m, 30° 58′ **2.** 36.16 m **3.** 37° 14′
4. 27.92 m **5.** 19° 10′ 40″ **6.** 53° 8′, 6 m

7. 81.9 mm **8.** 15.05 m **9.** 37.4 m
10. 113 m/s **11. a)** 2 **b)** $\sqrt{12}$ **c)** $\sqrt{12}$ **d)** $\sqrt{24}$ **e)** $\frac{1}{2}$ **f)** 1

MENSURATION 6

1. a) 625 mm^2 **b)** 2625 mm^2 **c)** 1299 mm^2 **d)** 2464 mm^2 **e)** 2772 mm^2
2. 3.75 m^2 **3.** 156 mm **4.** 1112 mm^2
5. 707.1 mm **6.** 6257 mm^2 **7.** 4818 mm^2
8. 4.489×10^5 mm^2 **9.** 1450 mm^2 **10.** 2463 mm^2, 895.3 mm^2
11. 30 mm **12.** 15.01 mm
13. a) 1314 mm^2 **b)** 0.0183 m^2 **c)** 2.859×10^4 mm^2
14. 158.9 mm, 0.5415 m, 676.5 mm **15.** 41.83 m

MENSURATION 7

1. a) 1.97×10^5 mm^3 **b)** 4.666×10^4 mm^3 **c)** 4.909×10^4 mm^3
d) 3.61×10^5 mm^3 **e)** 5.237×10^5 mm^3
2. a) 5.895×10^4 mm^2 **b)** 2.2×10^4 mm^2 **3.** 9.45×10^7 mm^3
4. 1.5 m^3 **5.** 3.518 m^2 **6.** 0.8513 m^3
7. 253.4 mm **8.** 0.277 m^3 **9.** 0.27 m^3, 2.017 m^2
10. 35.786 mm **11.** 21.22 mm, 3.062×10^4 mm^2
12. 8.766×10^4 mm^3 **13.** 8.728×10^5 mm^3 **14.** 4136 mm^2

MENSURATION 8

1. 465.9 mm^2 **2.** 270.6 mm^2 **3.** 27.6%
4. 3.059×10^5 mm^2 **5.** 1.19 m
6. 127 mm *slant*, 119.7 mm *vertical*, 84.67 mm *dia.*
7. 487.6 mm **8.** 162.5 m **9.** 88.22 mm
10. 25.4 mm **11.** 319 mm, 2.6×10^5 mm^2
12. 62.84 mm^2

SELF TEST NO. 4

1. a) No answer **b)** 62.45 mm **c)** 1561 mm^2
2. a) 64° 24′ 53″ **b)** 27.7242° **c)** 1.5335 radians
3. a) 108 mm **b)** 377 mm **c)** 38° 26′
4. a) (i) 0.3979, 0.9174, 0.4338
(ii) 0.8410, 0.5411, 1.5542
b) 52° 33′, 41.36 mm, 68.02 mm
c) 1116.7 mm^2
5. 98 m **6. a)** 0.089 55 m^3 **b)** 2.0892 m^2
7. a) 5237 mm^3 **b)** 26.185 mm **8.** 90°, 60°, 30°
9. 3928 mm^2, 226.7 mm^2 **10.** 1.078 radians, 0.493 radians

ANSWERS TO GEOMETRY, TRIGONOMETRY AND MENSURATION

ANSWERS TO DIAGRAMMATIC REPRESENTATION

DIAGRAMMATIC 2

1. Diagram only **2.** Diagram only **3.** $L = 0.44W + 107$
4. $R = 0.03L$ **5.** $V = 0.2T + 55$ **6.** $F = \frac{9}{5}c + 32$
7. $V = 1.7F - 2$ **8.** $E = 1.5 - 0.6I$ **9.** $L = 100 + 0.002t$

SELF TEST NO. 3

1. 3.06 kg **2.** $x = 1.5$ **3.** 62N, Slope = 0.42
4. 3.3 secs 84 m/sec **5.** Diagram only **6.** Diagram only

ANSWERS TO DIAGRAMMATIC REPRESENTATION